IntechOpen

Supporting open minds since 2005

Time Series Analysis - Data, Methods, and Applications
http://dx.doi.org/10.5772/intechopen.78491
Edited by Chun-Kit Ngan

Contributors
Lamyaa Sadouk, Tong Shu, Yanyan Zheng, Shou Chen, Kin Keung Lai, Ming-Tao Chou, Chris Aldrich, Takashi Kuremoto, Masanao Obayashi, Shingo Mabu, Kunikazu Kobayashi, Rohan Raman, Basanta Das

First published in London, United Kingdom, 2019 by IntechOpen
IntechOpen is the global imprint of INTECHOPEN LIMITED, registered in England and Wales, registration number: 11086078, 7th floor, 10 Lower Thames Street, London, EC3R 6AF, United Kingdom
Printed in Croatia

British Library Cataloguing-in-Publication Data
A catalogue record for this book is available from the British Library

Additional hard and PDF copies can be obtained from orders@intechopen.com

Time Series Analysis - Data, Methods, and Applications
Edited by Chun-Kit Ngan
p. cm.
Print ISBN 978-1-78984-778-9
Online ISBN 978-1-78984-779-6
eBook (PDF) ISBN 978-1-78984-786-4

For EU product safety concerns:
IN TECH d.o.o., Prolaz Marije Krucifikse Kozulić 3, 51000 Rijeka, Croatia, info@intechopen.com or visit our website at intechopen.com.

Meet the editor

Dr. Chun-Kit Ngan has been an Assistant Teaching Professor of Data Science at Worcester Polytechnic Institute (WPI) since January 2018. Before joining WPI, he was an Assistant Professor of Information Science at the Pennsylvania State University-Great Valley (PSU-GV). His research interests are Decision Guidance and Support Systems (DGSS), including decision optimization models, deep/machine learning, computational algorithms, data analytics, and DGSS applications, to guide domain-specific decision makers to make better decisions and provide them with actionable recommendations. He received the Best Paper Award and the Best Student Paper Award at the 2013 and 2011 International Conference on Enterprise Information Systems. He was the recipient of the 2013-2014 Seed Money Grant and 2015-2016 Early Career Award for Research and Scholarship Excellence at PSU-GV.

Contents

Preface

A time series is a sequence of data points that are measured and ordered over uniform time intervals. Examples include: household energy consumption, heart rates, hourly air temperature measurements, daily company stock prices, monthly auto sales, and yearly sales figures. Time series analysis is a statistical technique that analyzes time series and then extracts meaningful statistics, characteristics, and insights from the data. Some applications include identifying cyclical changes in sales of a product, forecasting business budget based on historical trends, and studying employee turnover data and employee training data to determine if there is any dependence of employee training programs on employee turnover rates over time. Thus, analyzing time series data is an important topic that has gained significant interest in the past decade. However, due to the data pattern complexity in different domains and industries nowadays, time series analysis has become very challenging and this has inspired domain scientists and practitioners to develop and apply the updated methodologies and applications to solve real-world time series problems such as forecasting, classification, and feature extraction.

To address the above issues, the book editor has carefully selected a number of domain experts' papers that focus on data, methods, and applications on time series analysis that provide our readers with the latest information, developments, and trends in this research area. Specifically, this book is divided into three sections: (1) Mining complex patterns in time series data, (2) Deep neural networks for time series analytics, and (3) Time Series forecasting in real-world problems. Each section includes two chapters. Each chapter starts with motivational background, technical foundations, specific methodologies, and then ends with examples and case studies to explain the concepts and techniques that help our readers understand the content easily.

Section 1: Mining Complex Patterns in Time Series Data

- Process Fault Diagnosis for Continuous Dynamic Systems over Multivariate Time Series. Use grey level co-occurrence matrices and local binary patterns to extract features from time series for process fault diagnosis.

- Fuzzy Forecast Based on Fuzzy Time Series. Use fuzzy time series for interval prediction and long-term significance level analysis.

Section 2: Deep Neural Networks for Time Series Analytics

- Training Deep Neural Networks with Reinforcement Learning for Time Series Forecasting. Combine multiple restricted Boltzmann machines (RBMs) and multilayer perception (MLP) and use stochastic gradient ascent (SGA) training algorithm for time series forecasting.

- CNN Approaches for Time Series Classification. Use Convolutional Neural Network (CNN) models for time series classification.

Section 3: Time Series Forecasting in Real-world Problems

- Forecasting Shrimp and Fish Catch in Chilika Lake over Time Series Analysis. Develop and apply Seasonal Auto Regressive Integrated Moving Average (SARIMA) and SARIMA with regressors (SARIMAX) to predict shrimp and fish catch respectively in the Chilika Lake.

- Using Gray-Markov Model and Time Series Model to Predict Foreign Direct Investment Trend for Supporting China's Economic Development. Construct Gray-Markov model and time series model to forecast the trend of China's utilization of foreign direct investment.

On behalf of all the authors that contributed to this book, I would like to warmly thank Author Service Managers, Ms. Sara Debeuc and Ms. Sara Petanjek, as well as Senior Commissioning Editor, Ms. Ana Pantar, from IntechOpen, who have tirelessly contributed to and assisted me in the success of this publication.

Chun-Kit Ngan, Ph.D.
Assistant Teaching Professor,
Graduate Qualifying Project Coordinator,
Data Science Program,
Department of Computer Science,
Worcester Polytechnic Institute,
Worcester, MA, U.S.A

Section 1

Mining Complex Patterns in Time Series Data

Chapter 1

Process Fault Diagnosis for Continuous Dynamic Systems Over Multivariate Time Series

Chris Aldrich

Abstract

Fault diagnosis in continuous dynamic systems can be challenging, since the variables in these systems are typically characterized by autocorrelation, as well as time variant parameters, such as mean vectors, covariance matrices, and higher order statistics, which are not handled well by methods designed for steady state systems. In dynamic systems, steady state approaches are extended to deal with these problems, essentially through feature extraction designed to capture the process dynamics from the time series. In this chapter, recent advances in feature extraction from signals or multivariate time series are reviewed. These methods can subsequently be considered in a classical statistical monitoring framework, such as used for steady state systems. In addition, an extension of nonlinear signal processing based on the use of unthresholded or global recurrence quantification analysis is discussed, where two multivariate image methods based on gray level co-occurrence matrices and local binary patterns are used to extract features from time series. When considering the well-known simulated Tennessee Eastman process in chemical engineering, it is shown that time series features obtained with this approach can be an effective means of discriminating between different fault conditions in the system. The approach provides a general framework that can be extended in multiple ways to time series analysis.

Keywords: process fault diagnosis, statistical process control, machine learning, time series analysis, deep learning

1. Introduction

In the process industries, advanced process control is widely recognized as essential to meet the challenges arising from the trend toward more complex, larger scale circuit configurations, plant-wide integration, and having to make do with fewer personnel. In these environments, characterized by highly automated process operations, algorithms to detect and classify abnormal trends in process measurements are critically important.

Process diagnostic algorithms can be derived from a continuum spanning first principle models on one end to entirely data driven or statistical models on the other. The latter is typically based on historical process data and is seen as the most cost effective approach to deal with complex systems. As a consequence, diagnostic methods have seen considerable growth over the last couple of decades.

Data-driven fault diagnosis can be traced back to control charts invented by Walter Shewhart at Bell Laboratories in the 1920s to improve the reliability of their telephony transmission systems. In these statistical process control charts, variables of interest were plotted as time series within statistical upper and lower limits. Shewhart's methodology was subsequently popularized by Deming and these statistical concepts, such as Shewhart control charts (1931), cumulative sum charts (1954), and exponentially weighted moving average charts were well established by the 1960s [1].

These univariate control charts do not exploit the correlation that may exist between process variables. In the case of process data, crosscorrelation is present, owing to restrictions enforced by mass and energy conservation principles, as well as the possible existence of a large number of different sensor readings on essentially the same process variable. These shortcomings have given rise to multivariate methods or multivariate statistical process control and related methods that have proliferated exponentially over the last number of years. These approaches can be viewed on the basis of the elementary operations involved in the fault diagnostic process, as outlined in **Figure 1** [2].

In this diagram, (i) a data matrix ($\check{X}$), representative of the process, is preprocessed or transformed to (ii) data matrix $\mathbf{X}$ and then mapped to (iii) a feature space ($\mathbf{F}$) within some bounded region (iv) L_F. These features can be used to (v) reconstruct the data ($\hat{\mathbf{X}}$), from which (vi) an error matrix (E) is generated, with scores again mostly confined to some bounded region L_E (vii).

Fault detection and fault diagnosis are typically done in both the feature space ($\mathbf{F}$) and the error space (E), based on the use of forward ($\mathfrak{I}$) and reverse mapping ($\mathfrak{R}$) models and suitable confidence limits L_F and L_E for the feature and error spaces. Alternatively, forward mapping in to the feature space can be only used for process monitoring.

Preprocessing of the data prior to fault diagnosis has received considerable attention over the last decade or so as a basis for the development of methods that can deal with nonlinearities in the data, lagged variables and unfolding of higher dimensional data. These approaches will mostly be discussed in the second part of the chapter.

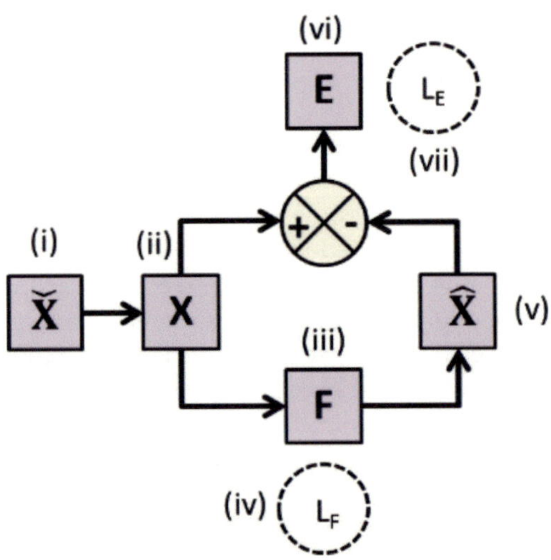

Figure 1.
Generalized framework for unsupervised process fault diagnosis.

1.1 Steady state systems

Linear steady state Gaussian processes and the use of principal component analysis will first be considered as an example on the basis of this general framework, after which other methods proposed over the last few decades will be reviewed.

As mentioned in the previous section, univariate control charts do not exploit the correlation that may exist between process variables and when the assumptions of linearity, steady state, and Gaussian behavior hold, multivariate statistical process control based on the use of principal component analysis can be used very effectively for early detection and analysis of any abnormal plant behavior. Since principal component analysis plays such a major role in the design of these diagnostic models, a brief outline of the methodology is in order.

Analysis, monitoring, and diagnosis of process operating performance based on the use of principal components is well established. The basic theory can be summarized as follows, where $\mathbf{X} \in \mathbb{R}^{N\mathrm{x}M}$ comprises the data matrix representative of the process with M variables and N observations, $\mathbf{S}$ is the covariance matrix of the process variables typically scaled to zero mean and unit variance, $\mathbf{P}$ is the loading matrix of the first $k < M$ principal components, $\mathbf{\Lambda}$ is a diagonal matrix containing the k eigenvalues of the decomposition, $\widetilde{\mathbf{P}}$ is the loading matrix of the $M - k$ remaining principal components, and $\widetilde{\mathbf{\Lambda}}$ is a diagonal matrix containing the $M - k$ remaining eigenvalues of the decomposition. The T^2 and Q-diagnostics (Eqs. 2 and 3) are commonly used in process monitoring schemes.

$$\mathbf{S} = \frac{\mathbf{X}^{\mathrm{T}}\mathbf{X}}{N-1} = \mathbf{P\Lambda P}^{\mathrm{T}} + \widetilde{\mathbf{P}}\widetilde{\mathbf{\Lambda}}\widetilde{\mathbf{P}}\mathbf{T} \tag{1}$$

$$\mathbf{Q} = (\mathbf{x} - \hat{\mathbf{x}})^{\mathrm{T}}(\mathbf{x} - \hat{\mathbf{x}}) = \mathbf{x}^{\mathrm{T}}\mathbf{Cx}, \text{ where } \mathbf{C} = \mathbf{P}\widetilde{\mathbf{P}}\mathbf{T} \tag{2}$$

$$\mathbf{T}^2 = \mathbf{t}^{\mathrm{T}}\mathbf{\Lambda}^{-1}\mathbf{t} = \mathbf{x}^{\mathrm{T}}\mathbf{Dx}, \text{ where } \mathbf{D} = \mathbf{P\Lambda}^{-1}\mathbf{P}^{\mathrm{T}} \tag{3}$$

In classical multivariate statistical process control based on principal component analysis, the control limits required for automated process monitoring are based on the assumption that the data are normally distributed. The α upper control limit for T^2 is calculated from N observations based on the F-distribution, that is,

$$UCL_{T^2(PCA)} = \frac{k(N+1)(N-1)F_{\alpha,k,N-k}}{N(N-k)} \tag{4}$$

Then upper control limit for Q is calculated by means of the χ^2 distribution as:

$$UCL_{Q(PCA)} = \frac{\Lambda_1\left[1 + c_\alpha\sqrt{(2\Lambda_2\theta^2)}/\Lambda_1 + \Lambda_2\theta(\theta-1)/\theta{\Lambda_1}^2\right]}{\theta} \tag{5}$$

where $\Lambda_1 = \sum_{k+1}^{M}\lambda_{ji}$ (for i = 1, 2, 3) and $\theta = 1 - 2\Lambda_1\Lambda_3/3\Lambda_2^2$. The standard normal deviates, c_α corresponding to the upper (1−α) percentile, while M is the total number of principal components (variables). The residual Q_i is more likely to have a normal distribution than the principal component scores, since it is a measure of the nondeterministic behavior of the system.

1.2 Unsteady state systems

Unlike steady state systems, unsteady state or dynamic systems show time dependence. This time dependence implies the presence of autocorrelation and/or nonstationarity [3]. Autocorrelation arises when the observations within a time

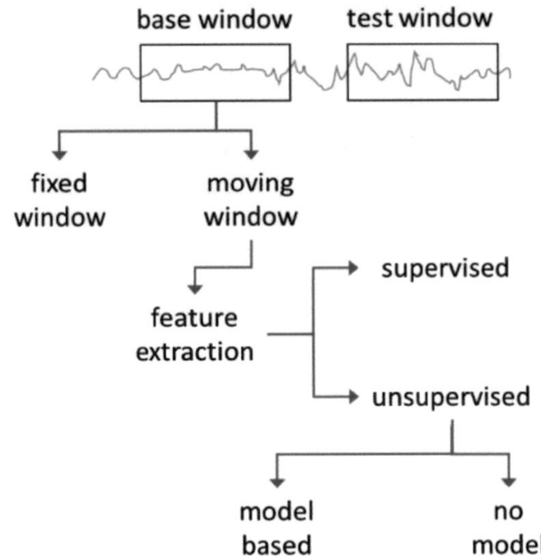

Figure 2.
Dynamic process monitoring as an extension of steady state approaches.

series are not independent, while nonstationarity means that the parameters governing a process change over time, for example, the mean, covariance or other higher order statistics. Therefore, in principle at least, these systems cannot be treated directly by the methods dealing with steady state systems.

Broadly speaking, methodologies dealing with dynamic process systems are all aimed at dealing with the issues arising from the time dependence of the data. Essentially, these approaches are based on the analysis of a segment of the time series data, as captured by a fixed or a moving window, as indicated in **Figure 2**. The time series segment amounts to observation of the process over a time interval, and the window length should be sufficient to capture the dynamics of the systems.

Dynamic process monitoring can be as simple as monitoring the mean or the variance of a signal, in which case, a test window as shown in **Figure 2** would not be required, and model maintenance would not be an issue. In more complex systems, as could be characterized by large multivariate sets of signals or high-dimensional signals, such as streaming video or hyperspectral data, feature extraction is often model-based. That is, a model derived from the data in the base window is applied to the data in the test window. For example, principal component models can be used for this purpose.

Where models are used and the nature of the signals changes as a result of process drift, recalibration of the models need to be done either at regular intervals or episodically, that is, when a change occurs. Some models, such as those based on principal and independent components can be updated recursively, as discussed in more detail in Sections 4 and 5. Alternatively, the model is updated *ab initio* at regular intervals.

Moreover, most feature extraction methods are unsupervised, that is, the time series data are unlabeled. Where supervised methods are used, features are extracted based on their ability to predict some label, such as the future evolution of the time series.

2. Unsupervised feature extraction

In principle, any low-dimensional representation of the time series data would constitute a feature set, that is, the data in the time series window, $\mathbf{X} \in \mathbb{R}^{N \times M}$ containing N measurements of the M plant variables with time lagged copies of these variables. These features can subsequently be dealt with by the same methods used for steady state systems, such as principal component analysis, independent component analysis, kernel methods, etc., some of which are considered in more detail below.

2.1 Dynamic principal component analysis (DPCA)

In dynamic PCA, first proposed by Ku et al. [4], the PCA model is built on the data matrix **X** residing in the window, to account for auto- and crosscorrrelation between variables. This approach implicitly estimates the autoregressive structure of the data (e.g., [5]). As functions of the model, the T^2 and Q-statistics will also be functions of the lag parameters. Since the mean and covariance structures are assumed to be invariant, the same global model is used to evaluate observations at any future time point.

Although dynamic PCA is designed to deal with autocorrelation in the data, the resultant score variables will still be autocorrelated or even crosscorrelated when no autocorrelation is present [4, 6]. These autocorrelated score variables have the drawback that they can lead to higher rates of false alarms when using Hotelling's T^2 statistic.

Several remedies have been proposed to alleviate this problem, for example, wavelet filtering [7], ARMA filtering [6], and the use of residuals from predictive models [8]. Nonlinear PCA models have been considered by several authors [9–13].

2.2 Independent component analysis

Stefatos and Hamza [14] and Hsu et al. [15] have introduced diagnostic methods using an approach based on dynamic independent component analysis capable of accurately detecting and isolating the root causes of individual faults. Nonlinear variants of these approaches have been investigated by Cai et al. [16], who have integrated the kernel FastICA algorithm with a manifold learning method known as locality preserving projection. Moreover, kernel FastICA was used to integrate FastICA and kernel PCA to exploit the advantages of both algorithms, as indicated by Zhang and Qin [17], Zhang [18], and Zhang et al. [19].

2.3 Slow feature analysis

Slow feature analysis [20] is an unsupervised learning method, whereby functions $\boldsymbol{g}(\boldsymbol{x})$ are identified to extract slowly varying features $\boldsymbol{y}(t)$ from rapidly varying signals $\boldsymbol{x}(t)$. This is done virtually instantaneously, that is, one time slice of the output is based on very few time slices of the input. Extensions of the method have been proposed by other authors [21–23].

2.4 Multiscale methods

Multiscale methods can be seen as a complementary approach preceding feature extraction from the time series. In this case, each process variable is extended or replaced by different versions of the variable at different scales. For example, with multiscale PCA, wavelets are used to decompose the process variables under scrutiny into multiple scale representations before application of PCA to detect and identify faulty conditions in process operations. In this way, autocorrelation of variables is implicitly accounted for, resulting in a more sensitive method for detecting process anomalies. Multiscale PCA constitutes a promising extension of multivariate statistical process control methods, and several authors have reported successful applications thereof [24–27].

2.4.1 Wavelets

Bakshi [28, 29] has proposed the use of a nonlinear multiscale principal component analysis methodology for process monitoring and fault detection based on multilevel wavelet decomposition and nonlinear component extraction by the use of input-training neural networks. In this case, wavelets are first used to decompose the data into different scales, after which PCA was applied to the reconstituted time series data. Choi et al. [30] have proposed nonlinear multiscale multivariate monitoring of dynamic processes based on kernel PCA, while Xuemin and Xiaogang [31] have proposed an integrated multiscale approach where kernel PCA is used on measured process signals decomposed with wavelets and have also proposed a similarity factor to identify fault patterns.

2.4.2 Singular spectrum analysis

With SSA, the time series is first embedded into a p-dimensional space known as the *trajectory matrix*. Singular value decomposition is then applied to decompose the trajectory matrix into a sum of elementary matrices [32–34], each of which is associated with a process mode.

Subsequently, the elementary matrices that contribute to the norm of the original matrix are grouped, with each group giving an approximation of the original matrix. Finally, the smoothed approximations or modes of the time series are recovered by diagonal averaging of the elementary matrices obtained from decomposing the trajectory matrix. Although SSA is a linear method, it can readily be extended to nonlinear forms, such as kernel-based SSA or SSA with autoassociative neural networks. Nonetheless, it has not been used widely in statistical process monitoring as yet, although some studies have provided promising results [2, 35, 36].

Table 1 gives a summary of multiscale methods that have been considered in process monitoring schemes over the last two decades.

2.5 Phase space methods

Phase space methods rely on the embedding of the data in a so-called phase space, by the use of delayed vector methods, that is, $\mathbf{y} \in \mathbb{R}^{N\times 1} \rightarrow \mathbf{X} \in \mathbb{R}^{(N-m+1)\times m} = [\mathbf{x}(t), \mathbf{x}(t-k)\ldots\mathbf{x}(t-k(m-1))]$. Embedding can also be done by the use of principal components or singular value decomposition of $\mathbf{X} \in \mathbb{R}^{(N-m+1)\times m}$, where $k = 1$ and m is comparatively large. In the latter case, the scores of the eigenvectors would represent an orbit or attractor with some geometrical structure, depending on the frequencies with which different regions of the phase space are visited. The topology

Methodology	Comment	References
Wavelets	Variable decomposition with wavelets before building PCA models	[37–47]
Singular spectrum analysis	Different variants have been proposed	[2, 36, 48]

Table 1.
Data preprocessing methodologies for multiscale process monitoring.

of this attractor is a direct result of the underlying dynamics of the system being observed, and the changes in the topology are usually an indication of a change in the parameters or structure of the system dynamics. Therefore, descriptors of the attractor geometry can serve as sensitive diagnostic variables to monitor abnormal system behavior.

2.5.1 Phase space attractor descriptors

For process monitoring purposes, the data captured in a moving window are embedded in a phase space, and descriptors such as correlation dimension [49–51], Lyapunov exponents, and information entropy [49] have been proposed to monitor deterministic or potentially chaotic systems. These approaches have not found widespread adoption in the industry yet, since the reliability of the descriptors may be compromised by high levels of signal noise.

2.5.2 Complex networks

Process circuits or plants lend themselves naturally to representation by networks and process monitoring schemes can exploit this. For example, Cai et al. [52] have essentially considered a lagged trajectory matrix in the form of a complex network, whereby the variables and their lagged versions served as network vertices. The edges of the network were determined by means of kernel canonical correlation analysis (a nonlinear approach to correlation relationships between sets of variables). Features were extracted from the variables based on the dynamic average degree of each vertex in the network. A standard PCA model, as described in Section 1.1 was consequently used to monitor the process. Case studies have indicated that this could yield considerable improvement in the reliability of the model to detect process disturbances.

2.5.3 Local recurrence quantification analysis

Any given sequence of numbers or time series can be characterized by similarity matrix containing measures of similarity (e.g., Euclidean distances) between all pair-wise points in the time series. A recurrence matrix is generated by binary quantization of the similarity matrix, based on a user specified threshold value. This thresholded matrix can be portrayed graphically as a recurrence plot, amenable to qualitative interpretation. The recurrence matrix, consisting of zeros and ones, can also be used as a basis to extract features that are representative of the dynamic behavior of the time series. This approach is widely referred to as recurrence quantification analysis, and in process engineering, it has mainly been used in the description of electrochemical phenomena and corrosion [53–58], but in principle has general applicability to any dynamic system.

2.5.4 Global recurrence quantification analysis

More recent extensions of recurrence quantification analysis have been considered by using the unthresholded similarity matrix as a basis for feature extraction. This is also referred to as global, as opposed to (local) recurrence quantification described in Section 2.5.3. The resulting recurrence plot can consequently be treated as an artificial image amenable to analysis by a large variety of algorithms normally applied to textural images, as discussed in more detail in Section 4.

3. Supervised feature extraction

3.1 Autoregressive models

Autocorrelated data can be addressed by fitting models to the data and analyzing the residuals, instead of the variables. With ARIMA models, crosscorrelation between the variables is not accounted for, and although multivariate models can also be employed using this approach, it becomes a complex task when there are many variables ($m > 10$), owing to the high number of parameters that must be estimated, as well as the presence of crosscorrelation [3, 59].

Apart from ARIMA models, other models, such as neural networks [60–62], decision trees [63], and just-in-time-learning with PCA [64], have also been proposed.

3.2 State space models

If it is assumed that the data matrix **X** contains all the dynamic information of the system, then the use of predictive models can be viewed as an attempt to remove all the dynamic information from the system to yield Gaussian residuals that can be monitored in the normal way. State space models offer a principled approach for the identification of the subspaces containing the data. This can be summarized as follows

$$\mathbf{x}_{k+1} = f(\mathbf{x}_k) + \mathbf{w}_k \tag{6}$$

$$\mathbf{y}_k = g(\mathbf{x}_k) + \mathbf{v}_k \tag{7}$$

Class	Method		References
Unsupervised feature extraction	Dynamic PCA	Linear PCA	[4, 8, 78, 79]
		Partial PCA	[12, 13]
		Kernel PCA	[80]
		Multiscale	[81]
	Dynamic ICA	ICA	[14, 82–87]
		Kernel ICA	[88]
	Slow feature analysis		[22–24]
	ICA	Standard	[89]
		Kernel	[88]
	Phase space and related methods	Attractor descriptors	[49–51, 90]
		Recurrence quantification analysis	[2, 53–58]
		Complex networks	[52]
	Dissimilarity		[84, 91, 92]
Supervised feature extraction	Autoregressive models		[59, 71, 93]
	State space models		[65–70, 76, 77, 93–95]
	Machine learning	Conventional	[2, 48, 60–62]
		Deep learning	[76, 77, 96]

Table 2.
Approaches to the monitoring of continuous dynamic process systems.

$\mathbf{x}_k$ and $\mathbf{y}_k$ are the respective state and measurement vectors of the system, and $\mathbf{w}_k$ and $\mathbf{v}_k$ are the plant disturbances and measurement errors, respectively, at time k. State space models and their variants have been considered by several authors [65–75].

3.3 Machine learning models

In principle, machine learning models are better able to deal with complex nonlinear systems than linear models, and some authors have considered the use of these approaches. For example, Chen and Liao [62] have used a multilayer perceptron neural network to remove the nonlinear and dynamic characteristics of processes to generate residuals that could be used as input to a PCA model for the construction of simple monitoring charts. Guh and Shiue [63] have used a decision tree to detect shifts in the multivariate means of process data. Auret and Aldrich [48] have considered the use of random forests in the detection of change points in process systems. In addition, Aldrich and Auret [2] have compared the use of random forests with autoassociative neural networks and singular spectrum analysis in a conventional process monitoring framework.

The application of deep learning in process monitoring is an emerging area of research that shows particular promising. This includes the use of stacked autoencoders [76], deep long short term memory (LSTM) neural networks [77], and convolutional neural networks. **Table 2** gives an overview of the feature extraction methods that have been investigated over the last few decades.

4. Case study: Tennessee Eastman process

Finally, as an example of the application of a process monitoring scheme incorporating feature extraction from time series data in a moving window, the following study can be considered. It is based on the Tennessee Eastman benchmark process widely used in these types of studies. The feature extraction process considered here is an extension of recurrent quantitative analysis discussed in Section 2.5.2. Instead of using thresholded recurrence plots, unthresholded or global recurrence plots are considered, as explained in more detail in below.

4.1 Tennessee Eastman process data

The Tennessee Eastman (TE) process as proposed by Downs and Vogel [97] and has been used as a benchmark in numerous process control and monitoring studies [98]. It captures the dynamic behavior of an actual chemical process, the layout of which is shown in **Figure 3**.

The plant consists of 5 units, namely a reactor, condenser, compressor, stripper and separator, as well as eight components (four gaseous reactants A, C, D, E, one inert reactant B, and three liquid products F, G, H) [97]. In this instance, the plant-wide control structure suggested by Lyman and Georgakis [99] was used to simulate the process and to generate data related to varying operating conditions. The data set is available at http://web.mit.edu/braatzgroup.

A total of four data sets were used, that is, one data set associated with NOC and the remaining three associated with three different faults conditions. The TE process comprises 52 variables, of which 22 are continuous process measurements, 19 are composition measurements, and the remaining 11 are manipulated variables. These variables are presented in **Table 3**. Each data set consisted of 960 measurements sampled at 3 min intervals.

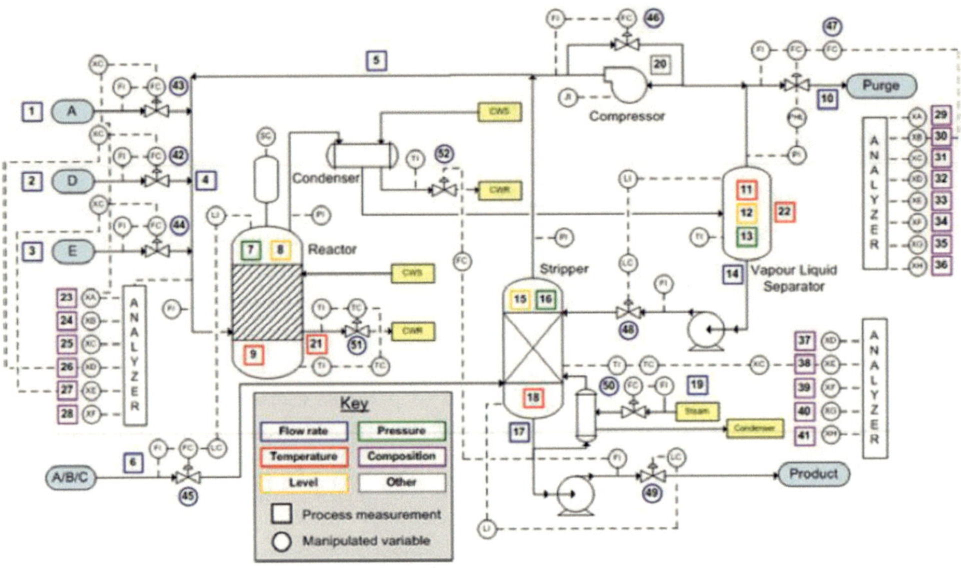

Figure 3.
Process flow of Tennessee Eastman benchmark problem.

The NOC samples were used to construct an off-line process monitoring model that consisted of a moving window of length b, moving sliding along the time series with a step size s. The three fault conditions are summarized in **Table 4**. Fault conditions 3, 9, and 15 are the most difficult to detect, and many fault diagnostic approaches fail to do so reliably.

In this case study, the approach previously proposed by Bardinas et al. [96] is applied to the three fault conditions in the TE process. The methodology can be briefly summarized as shown in **Figure 4**.

A window of user defined length b slides along the time series (A) with a user defined step size s, yielding time series segments (B), each of which can be represented by a similarity matrix (C) that is subsequently considered as an image from which features can be extracted via algorithms normally used in multivariate image analysis (D).

4.2 Feature extraction

Two sets of features were extracted from the similarity or distance matrices, namely features from the gray level co-occurrence matrices of the images, as well as local binary pattern features, as briefly discussed below.

4.2.1 Gray level co-occurrence matrices (GLCMs)

GLCMs assign distributions of gray level pairs of neighboring pixels in an image based on the spatial relationships between the pixels. More formally, if $y(i,j)$ is an element of a GLCM associated with an image I of size $R \times S$, having L gray levels, then $y(i,j)$ can be defined as

$$y(i,j) = \sum_{r=1}^{R} \sum_{s=1}^{S} \begin{cases} 1, & \textit{if } I(r,s) = i, \textit{and } I(r + \Delta r, s + \Delta s) = j \\ 0, & \textit{otherwise} \end{cases} \tag{8}$$

Process measurement		Composition measurement		Manipulated variable	
Variable	**Description**	**Variable**	**Description**	**Variable**	**Description**
1	A Feed	23	Reactor feed component A	42	D feed flow
2	D Feed	24	Reactor feed component B	43	E feed flow
3	E Feed	25	Reactor feed component C	44	A feed flow
4	Total Feed	26	Reactor feed component D	45	Total feed flow
5	Recycle flow	27	Reactor feed component E	46	Compressor recycle valve
6	Reactor feed rate	28	Reactor feed component F	47	Purge valve
7	Reactor pressure	29	Purge component A	48	Separator product liquid flow
8	Reactor level	30	Purge component B	49	Stripper product liquid flow
9	Reactor temperature	31	Purge component C	50	Stripper steam valve
10	Purge rate	32	Purge component D	51	Reactor cooling water flow
11	Separator temperature	33	Purge component E	52	Condenser cooling water flow
12	Separator level	34	Purge component F		
13	Separator pressure	35	Purge component G		
14	Separator underflow	36	Purge component H		
15	Stripper level	37	Product component D		
16	Stripper pressure	38	Product component E		
17	Stripper underflow	39	Product component F		
18	Stripper temperature	40	Product component G		
19	Stripper steam flow	41	Product component H		
20	Compressor work				
21	Reactor cooling water outlet temperature				
22	Separator cooling water outlet temperature				

Table 3.
Description of variables in Tennessee Eastman process.

Fault number	Description	Type
3	D feed temperature	Step change
9	Reactor feed D temperature	Random variation
15	Condenser cooling water valve	Sticking

Table 4.
Description faults 3, 9, and 15 in the Tennessee Eastman process.

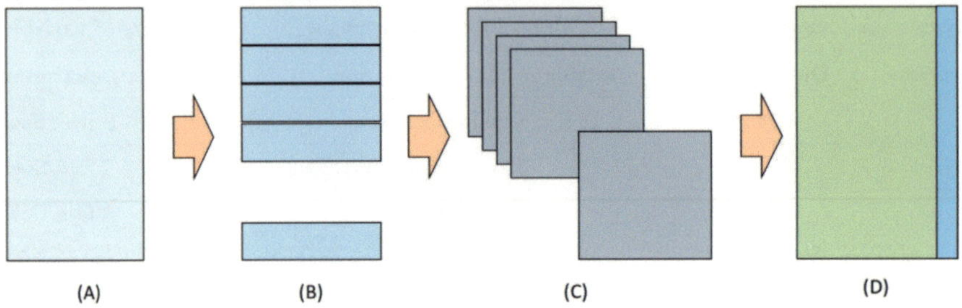

Figure 4.
Process monitoring methodology (after Bardinas et al., 2018). (A) Time series matrix, (B) Segmented time series matrix, (C) Distance matrices, and (D) Features and labels.

where (r, s) and $(r + \Delta r, s + \Delta s)$ denote the positions of the reference and neighboring pixels, respectively. From this matrix, various textural descriptors can be defined. Only four of these were used, as defined by Haralick et al. [100], namely contrast, correlation, energy, and homogeneity.

4.2.2 Local binary patters (LBPs)

LBPs are nonparametric descriptors of the local structure of the image [101]. The LBP operator is defined for a pixel in the image as a set of binary values obtained by comparing the center pixel intensity with its neighboring pixels. If the neighboring pixel exceeds the intensity of the center pixel value, this pixel is set to 1 (otherwise 0). Formally, given the central pixel's coordinates (x_0, y_0), the resulting LBP can be obtained in the decimal form as

$$LBP(x_0, y_0) = \sum_{p=0}^{p-1} s(i_p - i_0)2^p \tag{9}$$

where the gray level intensity value of the central pixel is i_0, that of its p'th neighbor is i_p. Moreover, the function $s(\bullet)$ is defined as

$$s(x) = \begin{cases} 0, & if\ x < 0 \\ 1, & if\ x \geq 0 \end{cases} \tag{10}$$

4.3 Selection of window length and step size

Apart from the selection of a feature extraction method, one of the main choices that need to be made in the process monitoring scheme is the length of the sliding window. If this is too small, the essential dynamics of the time series would not be captured. On the other hand, if it is too large, it would result in a considerable lag before any change in the process can be detected. There is also the possibility that transient changes may go undetected altogether. In the case of a moving window, the step size of the moves also needs to be considered. The selection of these two parameters can be done by means of a grid search, and the results of which are shown in **Figure 5**.

As indicated in **Figure 5**, the optimal window size was $b = 1000$ and the step size was $s = 20$ for both the GLCM and LBP features that were used as predictors. With these settings, the 500-tree random forest model [102] was able to differentiate between the normal operating conditions and the three fault classes with a reliability of approximately 82%.

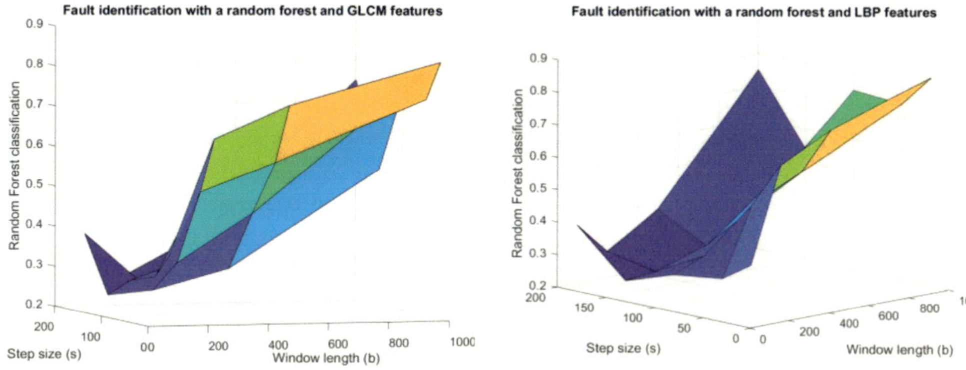

Figure 5.
Grid search optimization of the window length (b) and step size (s).

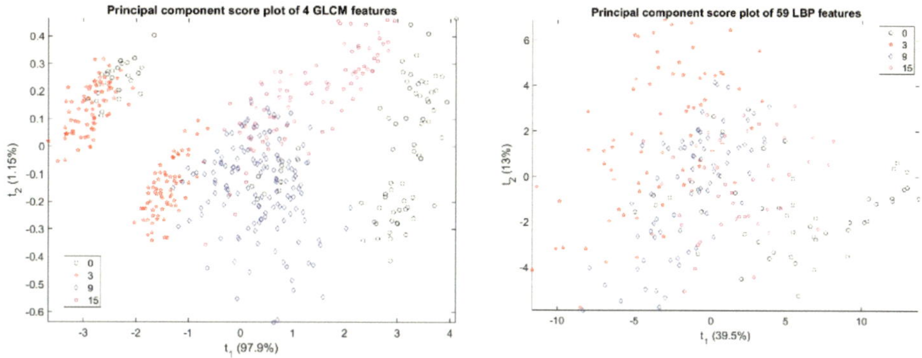

Figure 6.
Principal component score plots of GLCM (left) and LBP features (right).

In **Figure 6**, principal component score plots of the two optimal feature sets are shown. The large LBP feature set could not be visualized reliably, as the first two principal components could only capture 52.5% of the variance of the features. The variance in the smaller GLCM feature set could be captured with high reliability by the first two principal components of the four features. Here, the differences between the normal operating data ("0" legend) and the other fault conditions are clear ("3," "9," and "15").

4.4 Discussion

The approach outlined in Section 2.5.4. considered in more detail in the above case study is an extension of recurrent quantification analysis with the advantage that information is not lost when the similarity matrix of the signal is thresholded. Also, while thresholding does not preclude the use of a wide range of feature extraction algorithms, recurrent quantification has mostly been applied to dynamic systems based on a set of engineered features that allow some modicum of physical interpretation.

In most diagnostic systems, this is not essential and therefore more predictive feature sets may be constructed. These features could be engineered, as was considered in the case study or they could be learned, by taking advantage of state-of-the-art developments in deep learning.

In addition, the following general observations can be made not only with regard to the approach considered in this case study but also to other approaches reviewed in this chapter.

- Most of the nonlinear approaches used in steady state systems can be used in dynamic systems, and as a consequence, principal and independent component analysis and kernel methods have figured strongly in recent advances in dynamic process monitoring.

- With the routine acquisition of ever larger volumes of data and more complex processing, it can be expected that the field will continue to benefit from advances in machine learning. The application of deep learning methods in particular is a highly promising emerging area of research.

- Likewise, dynamic process monitoring is also likely to continue to benefit from closely related fields, such as process condition monitoring, structural health monitoring, change point detection, and novelty detection in other engineering or technical systems.

- As with steady state process monitoring, fault identification has received comparatively little attention to date.

5. Conclusions and future work

Data-driven fault diagnosis of dynamic systems has advanced considerably over the last decade or more. In this chapter, the large variety of algorithms currently available has been discussed in terms of a feature extraction problem associated with the data captured by sliding a window across the time series or in some cases making use of a fixed window. These features could be used in statistical process monitoring frameworks that are well established for steady state systems.

In addition, extension of a recent approach to nonlinear time series analysis, namely recurrence quantification analysis, has been considered and shown to be an effective means of monitoring dynamic process systems, such as represented by the Tennessee Eastman benchmark problem in chemical engineering.

As mentioned in Section 4.4., a wide range of feature extraction algorithms can be used with unthresholded or global recurrence quantification analysis. In future work, the application of convolutional neural networks to extract features from global recurrence plots will be considered. This does not necessarily require a large amount of data, as pretrained networks, such as AlexNet, ResNet, and VGG architectures, and others could possibly be used as is, in what would essentially be a texture analysis problem, similar to the work done by Fu and Aldrich [103, 104] in the recognition of flotation froth textures, for example.

Conflict of interest

The author declares no conflict of interest in this contribution.

Author details

Chris Aldrich
Western Australian School of Mines, Curtin University, Perth, WA, Australia

*Address all correspondence to: chris.aldrich@curtin.edu.au

References

[1] Russell EL, Chiang LH, Braatz RD. Data-Driven Techniques for Fault Detection and Diagnosis in Chemical Processes. London: Springer; 2000

[2] Aldrich C, Auret L. Unsupervised Process Monitoring and Fault Diagnosis with Machine Learning Methods. London: Springer-Verlag Ltd; 2013. Series: Advances in Pattern Recognition. ISBN: 978-1-4471-5184-5

[3] De Ketelaere B, Hubert M, Schmitt E. A review of PCA-based statistical process monitoring methods for time-dependent, high-dimensional data. (Downloaded from: https://wis.kuleuven.be/stat/robust/papers/2013/deketelaere-review.pdf on 26 December 2014). 2013

[4] Ku W, Storer RH, Georgakis C. Disturbance detection and isolation by dynamic principal component analysis. Chemometrics and Intelligent Laboratory Systems. 1995;**30**(1):179-196

[5] Tsung F. Statistical monitoring and diagnosis of automatic controlled processes using dynamic principal component analysis. International Journal of Production Research. 2000; **38**(3):625-637

[6] Kruger U, Zhou Y, Irwin GW. Improved principal component monitoring of large scale processes. Journal of Process Control. 2004;**14**(8): 879-888

[7] Luo R, Misra M, Himmelblau DM. Sensor fault detection via multiscale analysis and dynamic PCA. Industrial and Engineering Chemistry Research. 1999;**38**(4):1489-1495

[8] Rato TJ, Reis MS. Defining the structure of DPCA models and its impact on process monitoring and prediction activities. Chemometrics and Iintelligent Laboratory Systems. 2013; **125**:74-80

[9] Choi SW, Lee I-B. Nonlinear dynamic process monitoring based on dynamic kernel PCA. Chemical Engineering Science. 2004;**59**(24):5897-5908

[10] Cui P, Li J, Wang G. Improved kernel principal component analysis for fault detection. Expert Systems with Applications. 2008;**34**(2):1210-1219

[11] Shao J-D, Rong G. Nonlinear process monitoring based on maximum variance unfolding projections. Expert Systems with Applications. 2009;**36**(8): 11332-11340

[12] Li R, Rong G. Fault isolation by partial dynamic principal component analysis in dynamic process. Chinese Journal of Chemical Engineering. 2006a; **14**(4):486-493

[13] Li R, Rong G. Dynamic process fault isolation by partial DPCA. Chemical and Biochemical Engineering Quarterly. 2006b;**20**(1):69-77

[14] Stefatos G, Hamza AB. Dynamic independent component analysis approach for fault detection and diagnosis. Expert Systems with Applications. 2010;**37**(12):8606-8617

[15] Hsu C-C, Chen M-C, Chen L-S. A novel process monitoring approach with dynamic independent component analysis. Control Engineering Practice. 2010;**18**:242-253

[16] Cai L, Tian X, Zhang N. Non-Gaussian process fault detection method based on modified KICA. CIESC Journal. 2012;**63**(9):2864-2868. (In Chinese)

[17] Zhang Y, Qin SJ. Fault detection of nonlinear processes using multiway kernel independent component analysis. Industrial and Engineering Chemistry Research. 2007;**46**(23):7780-7787

[18] Zhang Y. Enhanced statistical analysis of nonlinear processes using KPCA, KICA and SVM. Chemical Engineering Science. 2009;**64**(5): 801-811

[19] Zhang Y, Li S, Teng Y. Dynamic process monitoring with recursive kernel principal component analysis. Chemical Engineering Science. 2012;**72**: 78-86

[20] Wiskott L, Berkes P, Franzius M, Sprekeler H, Wilbert N. Slow feature analysis. Scholarpedia. 2011;**6**(4):5282. Available from: http://www.scholarpedia.org/article/Slow_feature_analysis#WiskottSejnowski2002

[21] Zhang N, Tian X, Cai L, Deng X. Process fault detection based on dynamic kernel slow feature analysis. Computers and Electrical Engineering. 2015;**41**:9-17

[22] Shang C, Huang B, Yang F, Huang D. Slow feature analysis for monitoring and diagnosis of control performance. Journal of Process Control. 2016;**39**: 21-34

[23] Guo F, Shang C, Huang B, Wang K, Yang F, Huang D. Monitoring of operating point and process dynamics via probabilistic slow feature analysis. Chemometrics and Intelligent Laboratory Systems. 2016;**151**:115-125

[24] Fourie SH, De Vaal PL. Advanced process monitoring using an on-line non-linear multiscale principal component analysis methodology. Computers and Chemical Engineering. 2000;**24**(2–7):755-760

[25] Rosen C, Lennox JA. Multivariate and multiscale monitoring of wastewater treatment operation. Water Research. 2001;**35**(14):3402-3410

[26] Yoon S, MacGregor JF. Principal component analysis of multiscale data for process monitoring and fault diagnosis. AICHE Journal. 2004;**50**(11): 2891-2903

[27] Lee DS, Park JM, Vanrolleghem PA. Adaptive multiscale principal component analysis for on-line monitoring of a sequencing batch reactor. Journal of Biotechnology. 2005; **116**(2):195-210

[28] Bakshi BR. Multiscale PCA with application to multivariate statistical process monitoring. AICHE Journal. 1998;**44**:1596

[29] Bakshi BR. Multiscale analysis and modeling using wavelets. Journal of Chemometrics. 1999;**13**:415-434

[30] Choi SW, Morris AJ, Lee IB. Nonlinear multiscale modelling for fault detection and identification. Chemical Engineering Science. 2008;**63**(8): 2252-2266

[31] Xuemin T, Xiaogang D. A fault detection method using multi-scale kernel principal component analysis. In: Proceedings of the 27th Chinese Control Conference. Kunming, Yunnan, China; 2008

[32] Golyandina N, Nekrutkin V, Zhigljavsky A. Analysis of Time Series Structure: SSA and Related Techniques. New York, London: Chapman & Hall/CRC; 2001

[33] Jemwa GT, Aldrich C. Classification of process dynamics with Monte Carlo singular spectrum analysis. Computers and Chemical Engineering. 2006;**30**: 816-831

[34] Hassani H. Singular spectrum analysis: Methodology and comparison. Journal of Data Science. 2007;**5**:239-257

[35] Aldrich C, Jemwa GT, Krishnannair S. Multiscale process monitoring with singular spectrum analysis. Proceedings of the 12th IFAC Symposium on Automation in Mining, Mineral and

Metal Processing. Vol. 12(1). Quebec City, QC: Canada. 2007. pp. 167–172. Code 85804

[36] Krishnannair S, Aldrich C, Jemwa GT. Fault detection in process systems with singular spectrum analysis. Chemical Engineering Research and Design. 2016;**113**:151-168

[37] Alexander SM, Gor TB. Monitoring, diagnosis and control of industrial processes. Computers and Industrial Engineering. 1998;**35**(1–2): 193-196

[38] Kano M, Nagao K, Hasebe S, Hashimoto I, Ohno H, Strauss R, et al. Comparison of statistical process monitoring methods: Application to the Eastman challenge problem. Computers and Chemical Engineering. 2000;**24**: 175-181

[39] Misra M, Yue HH, Qin SJ, Ling C. Multivariate process monitoring and fault diagnosis by multiscale PCA. Computers and Chemical Engineering. 2002;**26**(9):1281-1293

[40] Li X, Yu Q, Wang J. Process monitoring based on wavelet packet principal component analysis. Computer Aided Chemical Engineering. 2003;**14**:455-460

[41] Ganesan R, Das T, Venkataraman V. Wavelet-based multiscale statistical process monitoring: A literature review. IIE Transactions. 2004;**36**:787-806

[42] Geng Z, Zhu Q. Multiscale nonlinear principal component analysis (nlpca) and its application for chemical process monitoring. Industrial and Engineering Chemistry Research. 2005;**44**(10): 3585-3593

[43] Wang D, Romagnoli JA. Robust multi-scale principal components analysis with applications to process monitoring. Journal of Process Control. 2005;**15**:869-882

[44] Maulud A, Wang D, Romagnoli JA. A multi-scale orthogonal nonlinear strategy for multi-variate statistical process monitoring. Journal of Process Control. 2006;**16**(7):671-683

[45] Zhang Y, Hu Z. Multivariate process monitoring and analysis based on multi-scale KPLS. Chemical Engineering Research and Design. 2011;**89**(12): 2667-2678

[46] Wang T, Liu X, Zhang Z. Characterization of chaotic multiscale features on the time series of melt index in industrial propylene polymerization system. Journal of the Franklin Institute. 2014;**351**:878-906

[47] Yang Y, Li X, Liu X, Chen X. Wavelet kernel entropy component analysis with application to industrial process monitoring. Neurocomputing. 2015;**147**(1):395-402

[48] Auret L, Aldrich C. Change point detection in time series data with random forests. Control Engineering Practice. 2010;**18**:990-1002

[49] Legat A, Dolecek V. Chaotic analysis of electrochemical noise measured on stainless steel. Journal of the Electrochemical Society. 1995; **142**(6):1851-1858

[50] Aldrich C, Qi BC, Botha PJ. Analysis of electrochemical noise with phase space methods. Minerals Engineering. 2006;**19**(14):1402-1409

[51] Xia D, Song S, Wang J, Shi J, Bi H, Gao Z. Determination of corrosion types from electrochemical noise by phase space reconstruction theory. Electrochemistry Communications. 2012;**15**(1):88-92

[52] Cai E, Liu D, Liang L, Xu G. Monitoring of chemical industrial processes using integrated complex network theory with PCA.

Chemometrics and Intelligent Laboratory Systems. 2015;**140**:22-35

[53] Cazares-Ibáñez E, Vázquez-Coutiño AG, García-Ochoa E. Application of recurrence plots as a new tool in the analysis of electrochemical oscillations of copper. Journal of Electroanalytical Chemistry. 2005;**583**(1):17-33

[54] Acun˜a-González, N, Garcia-Ochoa, E. and González-Sanchez, J. Assessment of the dynamics of corrosion fatigue crack initiation applying recurrence plots to the analysis of electrochemical noise data. International Journal of Fatigue. 2008;**30**:1211-1219

[55] Hou Y, Aldrich C, Lepkova K, Suarez LM, Kinsella B. Monitoring of carbon steel corrosion by use of electrochemical noise and recurrence quantification analysis. Corrosion Science. 2016;**112**:63-72

[56] Hou Y, Aldrich C, Lepkova K, Machuca LL, Kinsella B. Effect of electrode size on the electrochemical noise measured in different corrosion systems. Electrochimica Acta. 2017;**256**: 337-347

[57] Hou Y, Aldrich C, Lepkova K, Kinsella B. Detection of under deposit corrosion in CO2 environment by electrochemical noise and recurrence quantification analysis. Electrochimica Acta. 2018a;**274**:160-169

[58] Hou Y, Aldrich C, Lepkova K, Kinsella B. Identifying corrosion of carbon steel buried in iron ore and coal cargoes based on recurrence quantification analysis of electrochemical noise. Electrochimica Acta. 2018b;**283**:212-220

[59] Xie L, Zhang J, Wang S. Investigation of dynamic multivariate chemical process monitoring. Chinese Journal of Chemical Engineering. 2006; **14**(5):559-568

[60] Markou M, Singh S. Novelty detection: a review—Part 2: Neural network based approaches. Signal Processing. 2003;**83**(12):2499-2521

[61] Augusteijn MF, Folkert BA. Neural network classification and novelty detection. International Journal of Remote Sensing. 2002;**23**(14):2891-2902

[62] Chen J, Liao C-M. Dynamic process fault monitoring based on neural network and PCA. Journal of Process Control. 2002;**12**(2):277-289

[63] Guh R, Shiue Y. An effective application of decision tree learning for on-line detection of mean shifts in multivariate control charts. Computers and Industrial Engineering. 2008;**55**(2): 475-493

[64] Cheng C, Chiu M. Nonlinear process monitoring using JITL-PCA. Chemometrics and Intelligent Laboratory Systems. 2005;**76**:1-13

[65] Odiowei PP, Cao Y. State-space independent component analysis for nonlinear dynamic process monitoring. Chemometrics and Intelligent Laboratory Systems. 2010;**103**:59-65

[66] Odiowei PP, Cao Y. Nonlinear dynamic process monitoring using canonical variate analysis and kernel density estimations. IEEE Transactions on Industrial Informatics. 2009b;**6**(1): 36-45

[67] Simoglou A, Argyropoulos P, Martin EB, Scott K, Morris AJ, Taam WM. Dynamic modelling of the voltage response of direct methanol fuel cells and stacks part I: Model development and validation. Chemical Engineering Science. 2001;**56**:6761-6772

[68] Simoglou A, Martin EB, Morris AJ. Statistical performance monitoring of dynamic multivariate processes using state space modelling. Computers & Chemical Engineering. 2002;**26**:909-920

[69] Simoglou A, Georgieva P, Martin EB, Morris AJ, Feyo de Azevedo S. On-line monitoring of a sugar crystallization process. Computers & Chemical Engineering. 2005;**29**:1411-1422

[70] Russell EL, Chiang LH, Braatz RD. Fault detection in industrial processes using canonical variate analysis and dynamic principal component analysis. Chemometrics and Intelligent Laboratory Systems. 2000;**51**:81-93

[71] Negiz A, Cinar A. PLS, balanced, and canonical variate realization techniques for identifying VARMA models in state space. Chemometrics and Intelligent Laboratory Systems. 1997a;**38**(2):209-221

[72] Stubbs S, Zhang J, Morris AJ. Fault detection in dynamic processes using a simplified monitoring-specific CVA state space approach. Computer Aided Chemical Engineering. 2009;**26**:339-344

[73] Stubbs S, Zhang J, Morris AJ. Fault detection in dynamic processes using a simplified monitoring-specific CVA state space approach. Computers and Chemical Engineering. 2012;**41**:77-87

[74] Karoui MF, Alla H, Chatti A. Monitoring of dynamic processes by rectangular hybrid automata. Nonlinear Analysis: Hybrid Systems. 2010;**4**(4): 766-774

[75] Khediri IB, Limam M, Weihs C. Variable window adaptive kernel principal component analysis for nonlinear nonstationary process monitoring. Computers and Industrial Engineering. 2011;**61**(3):437-446

[76] Zhang Z, Jiang T, Li S, Yan Y. Automated feature learning for nonlinear process monitoring – An approach using stacked denoising autoencoder and k-nearest neighbor rule. Journal of Process Control. 2018; **64**:49-61

[77] Mehdiyev N, Lahann J, Emrich A, Enke D, Fettke P, Loos P. Time series classification using deep learning for process planning: A case from the process industry. Procedia Computer Science. 2017;**114**:242-249

[78] Lin WL, Qian Y, Li XX. Nonlinear dynamic principal component analysis for on-line process monitoring and diagnosis. Computers and Chemical Engineering. 2000;**24**(2–7):423-429

[79] Dobos L, Abonyi J. On-line detection of homogeneous operation ranges by dynamic principal component analysis based time-series segmentation. Chemical Engineering Science. 2012;**75**: 96-105

[80] Liu X, Krüger U, Littler TB, Xie L, Wang S. Moving window kernel PCA for adaptive monitoring of nonlinear processes. Chemometrics and Intelligent Laboratory Systems. 2009;**96**(2): 132-143

[81] Mele FD, Musulin E, Puigjaner L. Supply chain monitoring: A statistical approach. Computer Aided Chemical Engineering. 2005;**20**:1375-1380

[82] Lee JM, Yoo CK, Lee IB. Statistical monitoring of dynamic processes based on dynamic independent component analysis. Chemical Engineering Science. 2004;**59**:2995-3006

[83] Odiowei PP, Cao Y. Nonlinear dynamic process monitoring using canonical variate analysis and kernel density estimations. Computer Aided Chemical Engineering. 2009a;**27**: 1557-1562

[84] Rashid MM, Yu J. A new dissimilarity method integrating multidimensional mutual information and independent component analysis for non-Gaussian dynamic process monitoring. Chemometrics and Intelligent Laboratory Systems. 2012b; **115**:44-58

[85] Cai L, Tian X, Chen S. A process monitoring method based on noisy independent component analysis. Neurocomputing. 2014a;**127**:231-246

[86] Cai L, Tian X, Zhang N. A kernel time structure independent component analysis method for nonlinear process monitoring. Chinese Journal of Chemical Engineering. 2014b;**22** (11–12):1243-1253

[87] Cai L, Tian X. A new fault detection method for non-Gaussian process based on robust independent component analysis. Process Safety and Environmental Protection. 2014;**92**(6): 645-658

[88] Fan J, Wang Y. Fault detection and diagnosis of non-linear non-Gaussian dynamic processes using kernel dynamic independent component analysis. Information Sciences. 2014; **259**:369-379

[89] Chen J, Yu J, Mori J, Rashid MM, Hu G, Yu H, et al. A non-Gaussian pattern matching based dynamic process monitoring approach and its application to cryogenic air separation process. Computers & Chemical Engineering. 2013;**58**:40-53

[90] Ruschin-Rimini N, Ben-Gal I, Maimon O. Fractal geometry statistical process control for non-linear pattern-based processes. IIE Transactions. 2013; **45**(4):355-373

[91] Alabi S, Morris A, Martin E. On-line dynamic process monitoring using wavelet-based generic dissimilarity measure. Chemical Engineering Research and Design. 2005;**83**:698-705

[92] Yunus MYM, Zhang J. Multivariate process monitoring using classical multidimensional scaling and procrustes analysis. IFAC Proceedings Volumes (IFAC-PapersOnline). 2010;**9**(1): 165-170

[93] Negiz A, Cinar A. Statistical monitoring of multivariate dynamic processes with state-space models. AICHE Journal. 1997b;**43**(8):2002-2020

[94] Alawi A, Morris AJ, Martin EB. Statistical performance monitoring using state space modelling and wavelet analysis. In: Proceedings of the 15th European Symposium on Computer Aided Process Engineering. 2005. pp. 1375-1381

[95] Hill DJ, Minsker BS. Anomaly detection in streaming environmental sensor data: A data-driven modeling approach. Environmental Modelling and Software. 2010;**25**(9):1014-1022

[96] Bardinas JP, Aldrich C, Napier LFA. Predicting the operational states of grinding circuits by use of recurrence texture analysis of time series data. PRO. 2018;**6**:17

[97] Downs JJ, Vogel EF. A plant-wide industrial process control problem. Computers and Chemical Engineering. 1993;**17**(3):245-255

[98] Detroja KP, Gudi RD, Patwardhan SC. Fault detection using correspondence analysis: Application to Tennessee Eastman challenge problem. IFAC Proceedings Volumes. 2006;**39**(2): 705-710

[99] Lyman PR, Georgakis C. Plant-wide control of the Tennessee Eastman problem. Computers and Chemical Engineering. 1995;**19**(3):321-331

[100] Haralick RM, Shanmugam K, Dinstein IH. Textural features for image classification. IEEE Transactions on Systems, Man, and Cybernetics. 1973; **3**(6):610-621

[101] Ojala T, Pietikainen M, Harwood D. A comparative study of texture measures with classification based on featured distribution. Pattern Recognition. 1996;**29**(1):51-59

[102] Breiman L. Random forests. Machine Learning. 2001;**45**(1):5-32

[103] Fu Y, Aldrich C. Froth image analysis by use of transfer learning and convolutional neural networks. Minerals Engineering. 2018;**115**:68-78

[104] Fu Y, Aldrich C. Flotation froth image recognition with convolutional neural networks. Minerals Engineering. 2019;**132**:183-190

Chapter 2

Fuzzy Forecast Based on Fuzzy Time Series

Ming-Tao Chou

Abstract

This chapter mainly uses fuzzy time series for interval prediction and long-term significance level analysis. In this study, the Taiwan Shipping and Transportation Index (Taiwan STI) is used to illustrate the prediction process. Nine steps have been used to establish the interval prediction of the Taiwan Shipping and Transportation Index (Taiwan STI), and ΔS is called a long-term significance level (up/down/stable) is used to illustrate the long-term prediction significance level. By means of interval prediction and long-term prediction significance level, the future trends for this index and more internal messages related to this index can be provided to relevant researchers.

Keywords: fuzzy time series, interval prediction, long-term prediction significance level

1. Introduction

In 1965, Zadeh [1] proposed the concept of fuzzy sets as a tool to test the unknown degree of membership. Many fuzzy studies then attempted to use this method as a theoretical framework, which is widely used in the research fields of natural sciences and social sciences, obtaining good study achievements [2–22]. The fuzzy time series is also an analysis method derived from the concept of fuzzy sets. In 1993, Song and Chissom [18–21] successfully combined the concept of fuzzy sets with the time series model and began studies on fuzzy time series. Chen [3] proposed the simplified and easy-to-calculate method for Song and Chissom's model [18–21], so that the computation complexity of fuzzy time series is dramatically reduced. Lee and Chou [14] also proposed that rational settings of the lower and upper boundary in intervals of the universal set for fuzzy time series have improved their accuracy and reliability. Liaw [17] proposed a simple test method for whether a fuzzy time series has a fuzzy trend, in which the method is used to determine whether the data for analysis is in a steady state. Chou [12] added to Chen and Hsieh's defuzzification method [2] in the fuzzy time series, so that the long-term level for the series can be obtained, and the model originally used for single-point prediction can be applied to long-term prediction and interval prediction. This article mainly uses the algorithmic method of Chou's [12] research process for illustrating the fuzzy time series, taking the Taiwan Shipping and Transportation Index (STI) [23] as an example.

The remainder of this chapter is organized as follows. Section 2 presents the definition of fuzzy time series and Section 3 defines the long-term predictive significance level process. A numerical example of STI is shown in Section 4, and concluding remarks are mentioned in conclusion.

2. Definition of fuzzy time series

Fuzzy sets, presented by Zadeh [1], have numerous presentations, such as in fuzzy sets, fuzzy decision analysis, and fuzzy time series. The concept is also widely applied in social science article and applications [2–22]. Fuzzy time series is developed rapidly since their introduction by Song and Chissom [18–21]. Current fuzzy time series methods have benefited from both theoretical developments as well as relevant applications in research [2–22], which has led to more diverse uses. This trend indicates that the development of fuzzy time series has markedly improved. The definitions of the fuzzy time series used in this article are described as follows.

Definition 1 [18–21]. A fuzzy number on the real line $\mathfrak{R}$ is a fuzzy subset of $\mathfrak{R}$ that is normal and convex.

Definition 2 [18–21]. Let $Y(t)(t = ..., 0, 1, 2, ...)$, a subset of $\mathfrak{R}$, be the universe of discourse on which the fuzzy sets $f_i(t)(t = 1, 2, ...)$ are defined, and let $F(t)$ be the collection of $f_i(t)(t = 1, 2, ...)$. Then, $F(t)$ is called fuzzy time series on $Y(t)(t = ..., 0, 1, 2, ...)$.

Definition 3 [18–21]. Let I and J be the index sets for $F(t-1)$ and $F(t)$, respectively. If for any $f_j(t) \in F(t)$, where $j \in J$, there then exists $f_i(t-1) \in F(t-1)$, where $i \in I$, such that there exists a fuzzy relation $R_{ij}(t, t-1)$ and $f_j(t) = f_i(t-1) \circ R_{ij}(t, t-1)$, where '$\circ$' is the max–min composition. Then, $F(t)$ is said to be caused by only $F(t-1)$. Denote this as $f_i(t-1) \rightarrow f_j(t)$, or equivalently, $F(t-1) \rightarrow F(t)$.

Definition 4 [18–21]. If, for any $f_j(t-1) \in F(t)$, where $j \in J$, there exists $f_i(t-1) \in F(t-1)$, where $i \in I$, and a fuzzy relation $R_{ij}(t, t-1)$, such that $f_j(t) = f_i(t-1) \circ R_{ij}(t, t-1)$. Let $R(t, t-1) = \cup_{ij} R_{ij}(t, t-1)$, where $\cup$ is the union operator. Then, $R(t, t-1)$ is called the fuzzy relation between $F(t)$ and $F(t-1)$. Thus, we define this as the following fuzzy relational equation:

$$F(t) = F(t-1) \circ R(t, t-1).$$

Definition 5 [18–21]. Suppose that $R_1(t, t-1) = \cup_{ij} R^1{}_{ij}(t, t-1)$ and $R_2(t, t-1) = \cup_{ij} R^2{}_{ij}(t, t-1)$ are two fuzzy relations between $F(t)$ and $F(t-1)$. If, for any $f_j(t) \in F(t)$, where $j \in J$, there exists $f_i(t-1) \in F(t-1)$, where $i \in I$, and fuzzy relations $R^1{}_{ij}(t, t-1)$ and $R^2{}_{ij}(t, t-1)$ such that $f_i(t) = f_i(t-1) \circ R^1_{ij}(t, t-1)$ and $f_i(t) = f_i(t-1) \circ R^2_{ij}(t, t-1)$, then define $R_1(t, t-1) = R_2(t, t-1)$.

Definition 6 [18–21]. Suppose that $F(t)$ is only caused by $F(t-1)$, $F(t-2)$, ..., or $F(t-m)\ \ (m > 0)$. This relation can be expressed as the following fuzzy relational equation: $F(t) = F(t-1) \circ R_0(t, t-m)$, which is called the first-order model of $F(t)$.

Definition 7 [18–21]. Suppose that $F(t)$ is simultaneously caused by $F(t-1)$, $F(t-2)$, ..., and $F(t-m)\ \ (m > 0)$. This relation can be expressed as the following fuzzy relational equation: $F(t) = (F(t-1) \times F(t-2) \times ... \times F(t-m)) \circ R_a(t, t-m)$, which is called the mth-order model of $F(t)$.

Definition 8 [3]. $F(t)$ is fuzzy time series if $F(t)$ is a fuzzy set. The transition is denoted as $F(t-1) \rightarrow F(t)$.

Definition 9 [7]. Let $d(t)$ be a set of real numbers: $d(t) \subseteq R$. We define an exponential function where

1. $y = \exp d(t) \Leftrightarrow \ln y = d(t)$ and

2. $\exp(\ln d(t)) = d(t)$, $\ln(\exp x) = d(t)$.

Definition 10 [14]. The universe of discourse $U = [D_L, D_U]$ is defined such that $D_L = D_{\min} - st_\alpha(n)/\sqrt{n}$ and $D_U = D_{\max} + st_\alpha(n)/\sqrt{n}$ when $n \le 30$ or $D_L = D_{\min} - \sigma Z_\alpha/\sqrt{n}$ and $D_U = D_{\max} + \sigma Z_\alpha/\sqrt{n}$ when $n > 30$, where $t_\alpha(n)$ is the $100(1-\alpha)$ percentile of the t distribution with n degrees of freedom. z_α is the $100(1-\alpha)$ percentile of the standard normal distribution. Briefly, if Z is an $N(0,1)$ distribution, then $P(Z \ge z_\alpha) = \alpha$.

Definition 11 [14]. Assuming that there are m linguistic values under consideration, let A_i be the fuzzy number that represents the ith linguistic value of the linguistic variable, where $1 \le i \le m$. The support of A_i is defined as follows:

$$\begin{cases} D_L + (i-1)\dfrac{D_U - D_L}{m}, \quad D_L + \dfrac{i(D_U - D_L)}{m}, 1 \le i \le m-1 \\ D_L + (i-1)\dfrac{D_U - D_L}{m}, \quad D_L + \dfrac{i(D_U - D_L)}{m}, i = m. \end{cases}$$

Definition 12 [17]. For a test H_0 : *nonfuzzy trend* against H_1 : *fuzzy trend*, where the critical region $C^* = \{C | C_2^k + C_2^{n-k} > C_\lambda = C_2^n \times (1-\lambda)\}$, the initial value of the significance level α is 0.2.

Definition 13 [8]. Let $d(t)$ be a set of real numbers $d(t) \subseteq R$. An upper interval for $d(t)$ is a number b such that $x \le b$ for all $x \in d(t)$. The set $d(t)$ is said to be an interval higher if $d(t)$ has an upper interval. A number, max, is the maximum of $d(t)$ if max is an upper interval for $d(t)$ and max $\in d(t)$.

Definition 14 [8]. Let $d(t) \subseteq R$. The least upper interval of $d(t)$ is a number $\overrightarrow{\max}$ satisfying:

1. $\overrightarrow{\max}$ is an upper interval for $d(t)$ such that $x \le \overrightarrow{\max}$ for all $x \in d(t)$ and

2. $\overrightarrow{\max}$ is the least upper interval for $d(t)$, that is, $x \le b$ for all $x \in d(t) \Rightarrow \overrightarrow{\max} \le b$.

Definition 15 [8]. Let $d(t)$ be a set of real numbers $d(t) \subseteq R$. A lower interval for $d(t)$ is a number b such that $x \ge b$ for all $x \in d(t)$. The set $d(t)$ is said to be an interval below if $d(t)$ has a lower interval. A number, min, is the minimum of $d(t)$ if min is a lower interval for $d(t)$ and min $\in d(t)$.

Definition 16 [8]. Let $d(t) \subseteq R$. The least lower interval of $d(t)$ is a number $\overleftarrow{\min}$ satisfying:

1. $\overleftarrow{\min}$ is a lower interval for $d(t)$ such that $x \ge \overleftarrow{\min}$ for all $x \in d(t)$ and.

2. $\overleftarrow{\min}$ is the least lower interval for $d(t)$, that is, $x \ge b$ for all $x \in d(t) \Rightarrow \overleftarrow{\min} \le b$.

Definition 17 [8]. The long-term predictive value interval $(\overleftarrow{\min}, \overrightarrow{\max})$ is called the static long-term predictive value interval.

Definition 18 [2]. Let $A_i = (\alpha_i, \ \beta_i, \ \gamma_i), i = 1, 2, \ldots, n,$ be n triangular fuzzy numbers. By using the graded mean integration representation (GMIR) method, the GMIR value $P(A_i)$ of A_i is $P(A_i) = (\alpha_i + 4\beta_i + \gamma_i)/6$. $P(A_i)$ and $P(A_j)$ are the GMIR values of the triangular fuzzy numbers A_i and A_j, respectively.

Definition 19 [12]. Set up new triangular fuzzy numbers by $S = (\overleftarrow{\min}, \hat{d}(t), \overrightarrow{\max})$. After GMIR transformation, S becomes a real number ΔS. This is called the long-term significance level with fuzzy time series. The ΔS is a real number satisfying the following:

1. ΔS is called a long-term significance level up, only if: $\Delta S > \hat{d}(t)$;

2. ΔS is called a long-term significance level down, only if: $\Delta S < \hat{d}(t)$; and

3. ΔS is called a long-term significance level stable, only if: $\Delta S = \hat{d}(t)$.

3. Procedure of fuzzy time series forecasting

This section proposes a method to forecast the long-term predictive significance level by Chou. The stepwise procedure of the proposed method consists the following steps [8], illustrated as a flowchart in **Figure 1** [5–12].

Step 1. Let $d(t)$ be the data under consideration and let $F(t)$ be fuzzy time series. Following Definition 11, a difference test is performed to determine whether stability of the information. Recursion is performed until the information is in a stable state, where the critical region is $C^* = \left\{C \middle| C_2^k + C_2^{m-k} > C_\lambda = C_2^m \times (1-\lambda)\right\}$.

Step 2. Determine the universe of discourse $U = [D_L, D_U]$.

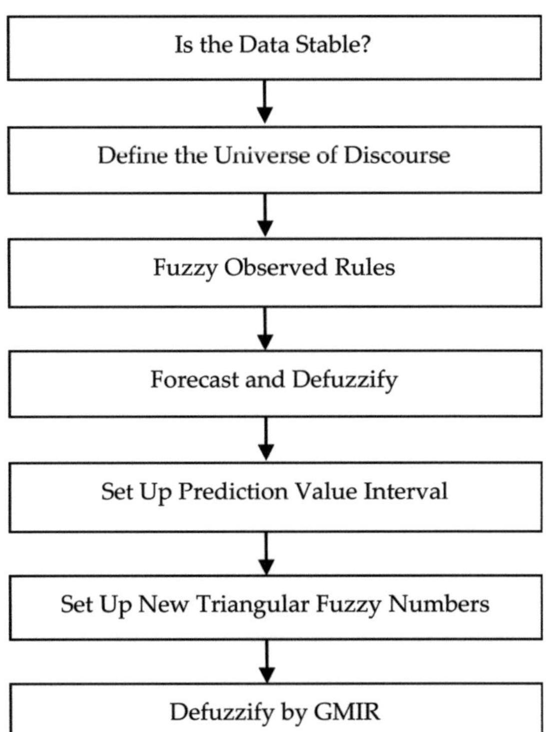

Figure 1.
Procedure of the proposed model.

Step 3. Define A_i by letting its membership function be as follows:

$$u_{A_i}(x) = \begin{cases} 1 \ \text{for}\ x \in \left[D_L + (i-1)\dfrac{D_U - D_L}{m}, D_L + \dfrac{i(D_U - D_L)}{m}\right) \\ \text{where } 1 \le i \le m-1; \\ 1 \ \text{for}\ x \in \left[D_L + (i-1)\dfrac{D_U - D_L}{m}, D_L + \dfrac{i(D_U - D_L)}{m}\right] \\ \text{where } i = m; \\ 0 \ \text{otherwise.} \end{cases}$$

Step 4. Then, $F(t) = A_i$ if $d(t) \in \text{supp}(A_i)$, where $\text{supp}(\cdot)$ denotes the support.

Step 5. Derive the transition rule from period $t-1$ to t and denote it as $F(t-1) \to F(t)$. Aggregate all transition rules. Let the set of rules be $R = \{r_i | r_i : P_i \to Q_i\}$.

Step 6. The value of $d(t)$ can be predicted using the fuzzy time series $F(t)$ as follows. Let $T(t) = \{r_j | d(t) \in \text{supp}(P_j), \textit{where } r_j \in R\}$ be the set of rules fired by $d(t)$, where $\text{supp}(P_j)$ is the support of P_j. Let $\overline{\text{supp}(P_j)}$ be the median of $\text{supp}(P_j)$. The predicted value of $d(t)$ is $\sum_{r_j \in T(t-1)} \frac{\overline{\text{supp}(Q_j)}}{|T(t-1)|}$.

Step 7. The long-term predictive value interval for $d(t)$ is given as $(\overleftarrow{\min}, \overrightarrow{\max})$.

Step 8. Set up new triangular fuzzy numbers by $\Delta S = (\overleftarrow{\min}, \hat{d}(t) \overrightarrow{\max})$.

Step 9. Defuzzify S to be ΔS.

4. Numerical example of Shipping and Transportation Index in Taiwan

In this study, the Shipping and Transportation Index (STI) in Taiwan is used for a numerical example. The STI reflects the spot rates of the Taiwan Stock Exchange Corporation. The STI data are sourced from the Taiwan Stock Exchange Corporation [23], the historical data for which is defined here as the STI, and month-averaged data for the period between January, 2015, and June, 2018, was collected.

Over these 42 data points, the analysis produces an average of 4.226, with a standard deviation of 0.172, maximum value of 4.571, and minimum value of 4.067. These descriptive statistics show that the STI has largely remained at the 1124.70 level. As shown in **Figure 2**, its current rate of return is negative.

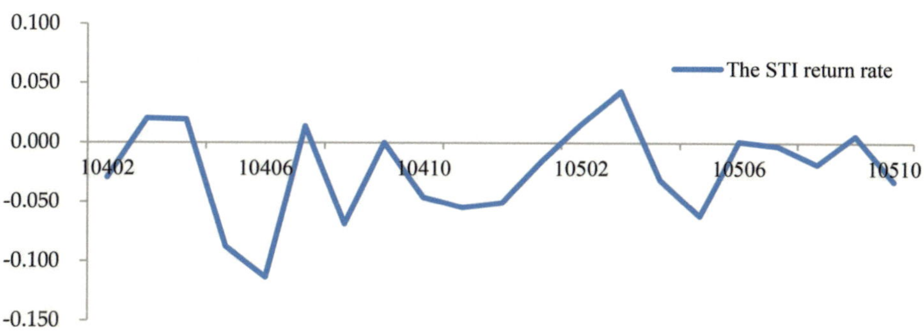

Figure 2.
Rate of return of the STI.

The following steps in the procedure are performed when using fuzzy time series to analyze STI.

Step 1. First, we take the logarithm of the STI data to reduce variation and improve the forecast accuracy, letting $STI(\tilde{t}) = \ln STI(t)$.

Step 2. Maintaining stationary data while forecasting helps to improve the forecast quality; therefore, we conduct a stationary test on the STI data. For fuzzy time series, a fuzzy trend test can measure whether the STI's fuzzy trend moves upward or downward. Using this fuzzy trend test, the STI data can be converted into a stationary series. If the original STI data exhibited a fuzzy trend, it can be eliminated by taking the difference. We then repeat the test after taking the first difference to measure if the STI data exhibits a fuzzy trend. If a fuzzy trend is again observed, then we take the second difference, and so on.

Letting $\mathrm{STI}(t)$ be the historical data under consideration and fuzzy time series, a difference test is used (following Definition 11) to understand whether the stability of the information. Recursion is performed until the information is determined to be stable. Once the region $C'' = \{C | C = C_2^{22} + C_2^{42-22}\} = 432 < \{C | C_2^{42} \times (1 - 0.2)\}$ $= 688.8$, the STI data are considered in a stable state and are not rejected.

Step 3. According to the interval setting of the STI data, we define the upper and lower bounds, which facilitate dividing the linguistic value intervals later. From Definition 10, the discourse $U = [D_L, D_U]$. From **Table 1**, $D_{\min} = 4.067$, $D_{\max} = 4.571$, $s = 0.172$, and $n = 42$ can be obtained. Letting$\alpha = 0.05$, since n is large than 30, a standard normal Z was used. Thus, $Z_{0.05} = 1.645$, $D_L = D_{\min} - st_\alpha/\sqrt{n} \approx 3.627$, and $D_U = D_{\max} + st_\alpha/\sqrt{n} \approx 5.011$. That is, $U = [3.627, 5.011]$.

Step 4. After defining the upper and lower bounds of the STI data in Step 3, we can define the SCFI range by determining the membership function as well as the linguistic values. We can also define the range of the subinterval for each linguistic value, assuming that the following linguistic values are under consideration: extremely few, very few, few, some, many, very many, and extremely many. According to Definition 11, the supports of fuzzy numbers that represent these linguistic values are given as follows:

$$u_{A_i}(x) = \begin{cases} 1 \ \text{ for } x \in [3.627 + (i-1)(0.129), 3.627 + i(0.198)) \\ \text{where } 1 \leq i \leq m-1; \\ 1 \ \text{ for } x \in [3.627 + (i-1)(0.129), 3.627 + i(0.198)] \\ \text{where } i = m; \\ 0 \ \text{ otherwise.} \end{cases}$$

where A_1 = "extremely few," A_2 = "very few," A_3 = "few," A_4 = "some," A_5 = "many," A_6 = "very many," and A_7 = "extremely many." Thus, the supports are $\mathrm{supp}(A_1) = [3.627, 3.825)$, $\mathrm{supp}(A_2) = [3.825, 4.023)$, $\mathrm{supp}(A_3) = [4.023, 4.221)$, $\mathrm{supp}(A_4) = [4.221, 4.419)$, $\mathrm{supp}(A_5) = [4.419, 4.617)$, $\mathrm{supp}(A_6) = [4.617, 4.815)$, and $\mathrm{supp}(A_7) = [4.815, 5.011]$.

Step 5. According to the subinterval setting of each linguistic value, we classified each historical dataset of the STI into its corresponding interval to measure the value corresponding to the linguistic value for each interval. The fuzzy time series $F(t)$ was given by $F(t) = A_i$ when $d(t) \in \mathrm{supp}(A_i)$. Therefore, $F(201501) = A_5$, $F(201502) = A_6$, $F(201503) = A_5$, $F(201504) = A_6$, ..., and $F(201806) = A_3$. **Table 1** shows the comparison between the actual SCFI data and the fuzzy enrollment data.

Step 6. We apply fuzzy theory to define the corresponding value for the intervals of the STI data, arrange the corresponding method for the STI data, and

Year	Actual	ln(Actual)	Fuzzified	The forecast value
201501	95.611	4.560	A_5	4.518
201502	92.839	4.531	A_5	4.518
201503	94.750	4.551	A_5	4.518
201504	96.622	4.571	A_6	4.617
201505	88.503	4.483	A_5	4.518
201506	79.003	4.369	A_4	4.221
201507	80.103	4.383	A_4	4.221
201508	74.787	4.315	A_4	4.221
201509	69.560	4.242	A_4	4.221
201510	71.416	4.269	A_4	4.221
201511	67.625	4.214	A_3	4.221
201512	64.282	4.163	A_3	4.221
201601	63.301	4.148	A_3	4.221
201602	64.315	4.164	A_3	4.221
201603	67.143	4.207	A_3	4.221
201604	65.073	4.176	A_3	4.221
201605	61.163	4.114	A_3	4.221
201606	61.221	4.114	A_3	4.221
201607	61.043	4.112	A_3	4.221
201608	59.942	4.093	A_3	4.221
201609	60.293	4.099	A_3	4.221
201610	58.372	4.067	A_3	4.221
201611	58.736	4.073	A_3	4.221
201612	57.892	4.059	A_3	4.221
201701	59.278	4.082	A_3	4.221
201702	62.746	4.139	A_3	4.221
201703	65.467	4.182	A_3	4.221
201704	62.142	4.129	A_3	4.221
201705	76.626	4.339	A_4	4.221
201706	63.029	4.144	A_3	4.221
201707	64.728	4.170	A_3	4.221
201708	68.464	4.226	A_4	4.221
201709	70.555	4.256	A_4	4.221
201710	67.830	4.217	A_4	4.221
201711	66.696	4.200	A_3	4.221
201712	67.640	4.214	A3	4.221
201801	69.769	4.245	A_4	4.221
201802	65.206	4.178	A_3	4.221
201803	64.669	4.169	A_3	4.221
201804	64.671	4.169	A_3	4.221

Year	Actual	ln(Actual)	Fuzzified	The forecast value
201805	63.759	4.155	A_3	4.221
201806	62.712	4.139	A_3	4.221

Table 1.
Fuzzy historical STI data and the forecasted results.

$r_1 : A_3 \to A_3$	$r_5 : A_5 \to A_4$
$r_2 : A_3 \to A_4$	$r_6 : A_5 \to A_5$
$r_3 : A_4 \to A_4$	$r_7 : A_5 \to A_6$
$r_4 : A_4 \to A_3$	$r_8 : A_6 \to A_5$

Table 2.
*Fuzzy transitions derived from **Table 1**.*

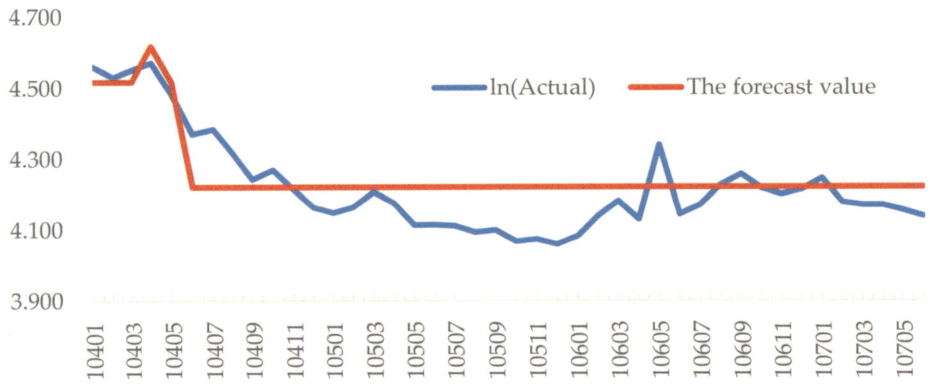

Figure 3.
Forecast STI and actual STI.

integrate the changes from all the rules to determine the rules for the STI. The transition rules are derived from **Table 1**. For example, $F(201501) \to F(201502)$ is $A_5 \to A_5$. **Table 2** shows all transition rules obtained from **Table 1**.

Step 7. We calculate each rule by determining all the rules of the STI, and the calculation results can be used to forecast future values. **Table 1** shows the forecasting results from 201001 to 201806.

Step 8. The calculated STI rules can define the intervals of the STI data; using these intervals, we can determine the variation in future long-term intervals. The long-term predictive value interval for the STI is given as (3.726, 4.913). Thus, the long-term predictive interval for the STI is given as (41.506, 136.022). Therefore, the current long-term S STI is bounded by this interval. According to Step 8, the fuzzy STI of 201501 shown in **Table 1** is A_5, and from **Table 2**, we can see that the rules are the fuzzy logical relationships in Rule 8 of **Table 2**, in which the current state of fuzzy logical relationships is A_3. Thus, the 201806 STI predictive value is 41.506.

Step 9. Letting defuzzified S be ΔS, the STI 201806 forecast value based on our investigation is 68.090, and its trading range is between 41.506 and 136.002. Thus, the new triangular fuzzy numbers by S = (41.506, 68.090, 136.002). Thus, the defuzzified S is ΔS = 74.981, and $\Delta S = 74.981 > \hat{d}(t) = 68.090$. ΔS is called a long-term significance level up.

The result shows that based on the long-term significance level, the STI is currently oversold. This result and the risk–reward ratio are both related within the group. We used **Table 1** data in our analysis according to the root mean square percentage error (R.M.S.P.E.) method, with an average prediction error of 1.708%. **Figure 3** shows the forecast visitor arrivals determined through fuzzy time series analysis and the actual STI values. Based on the fuzzy time series results, the average STI is estimated to be 68.090 in 201806 (**Figure 3**).

5. Conclusions and future work

In this article, a long-term predictive value interval model is developed for forecasting the STI. This model facilitates minimizing the uncertainties associated with fuzzy numbers. The method is examined by forecasting the STI by using data from which $\Delta S = 74.981$ and $\Delta S > \hat{d}(t)$ is obtained. For index returns, the current rate of return is negative and its volatility is increasing. The long-term predictive significance level of the STI is at the ΔS level; the STI should thus exhibit extreme volatility.

The current model for the STI 201806 forecast level deviates insignificantly from the actual values for an average of 68.090 and is within the group; the prediction error does not exceed 1.708% of the significance level. By constructing a fuzzy time series forecasting model for the STI with an error of less than 1.708%, with the traditional fuzzy time excluded from the single-point forecast comparison, this model provides a long-term predictive significance level.

Furthermore, the proposed method can be computerized. Thus, by improving fuzzy linguistic assessments as well as the evaluation of fuzzy time series, decision makers can automatically obtain the final long-term predictive significance level.

The STI used in this chapter is used as a forecasting example. If you predict that the future will rise, you can use the buying strategy. For example, if the index returns in the future, you can use the selling strategy.

The four functions of management are mainly four functions: planning, organization, leadership and control. The fuzzy time series mode used in this chapter can be applied to controlled projects to compare and correct whether the re-executed work meets expectations. If you meet expectations, re-plan the original settings.

Acknowledgements

This chapter is extended and revised the article "An improved fuzzy time series theory with applications in the Shanghai containerized freight index".

Author details

Ming-Tao Chou
Department of Aviation and Maritime Transportation Management, Chang Jung Christian University, Taiwan

*Address all correspondence to: mtchou@mail.cjcu.edu.tw

References

[1] Zadeh LA. Fuzzy set. Information and Control. 1965;**8**:338-353

[2] Chen S-H, Hsieh C-H. Representation, ranking, distance, and similarity of L-R type fuzzy number and application. Australian Journal of Intelligent Information Processing Systems. 2000;**6**:217-229

[3] Chen S-M. Forecasting enrollments based on fuzzy time series. Fuzzy Sets and Systems. 1996;**81**:311-319

[4] Chen S-M. Forecasting enrollments based on high order fuzzy time series. Cybernetics and Systems: An International Journal. 2002;**33**:1-16

[5] Chou M-T, Lee H-S. Increasing and decreasing with fuzzy time series. In: Joint Conference on Information Sciences; 2006. pp. 1240-1243

[6] Chou M-T. A fuzzy time series model to forecast the BDI. In: IEEE Proceedings of the Fourth International Conference on Networked Computing and Advanced Information Management; 2008. pp. 50-53

[7] Chou M-T. The logarithm function with a fuzzy time series. Journal of Convergence Information Technology. 2009;**4**:47-51

[8] Chou M-T. Long-term predictive value interval with the fuzzy time series. Journal of Marine Science and Technology. 2011;**19**:509-513

[9] Chou M-T. An application of fuzzy time series: A long range forecasting method in the global steel price index forecast. Review of Economics and Finance. 2013;**3**:90-98

[10] Chou M-T, Chou C-C. The implication of Taiwan's ore tramp carrier cargo on the blast furnace plant. Advanced Materials Research. 2013;**694–697**:3488-3491

[11] Chou MT. Fuzzy time series theory application for the China containerized freight index. Applied Economics and Finance. 2016;**3**:444-453

[12] Chou MT. An improved fuzzy time series theory with applications in the Shanghai containerized freight index. Journal of Marine Science and Technology. 2017;**25**:393-398

[13] Duru O, Yoshida S. Modeling principles in fuzzy time series forecasting. In: 2012 IEEE Conference on Computational Intelligence for Financial Engineering and Economics; 2012. pp. 1-7

[14] Lee H-S, Chou M-T. Fuzzy forecast based on fuzzy time series. International Journal of Computer Mathematics. 2004;**81**:781-789

[15] Lai RK, Fan CY, Huang WH, Chang PC. Evolving and clustering fuzzy decision tree for financial time series data forecasting. Expert Systems with Applications. 2009;**36**:3761-3773

[16] Liang M-T, Wu J-H, Liang G-S. Applying fuzzy mathematics to evaluating the membership of existing reinforced concrete bridges in Taipei. Journal of Marine Science and Technology. 2006;**8**:16-29

[17] Liaw M-C. The order identification of fuzzy time series, models construction and forecasting [PhD thesis]. Taiwan: National Chengchi University; 1997

[18] Song Q, Chissom BS. Forecasting enrollment with fuzzy time series—Part I. Fuzzy Sets and Systems. 1993;**54**: 1-9

[19] Song Q, Chissom BS. Fuzzy time series and its models. Fuzzy Sets and Systems. 1993;**54**:269-277

[20] Song Q, Chissom BS. Forecasting enrollment with fuzzy time series—Part II. Fuzzy Sets and Systems. 1994;**62**:1-8

[21] Song Q, Leland RP, Chissom BS. Fuzzy stochastic time series and its models. Fuzzy Sets and Systems. 1997; **88**:333-341

[22] Teoh HJ, Chen CH, Chu HH, Chen JS. Fuzzy time series model based on probabilistic approach and rough set rule induction for empirical research in stock markets. Data & Knowledge Engineering. 2008;**67**:103-117

[23] Taiwan Stock Exchange Corporation [Internet]. 2018. Available from: http://www.twse.com.tw/en/ [Accessed: August 03, 2018]

Section 2

Deep Neural Networks for Time Series Analytics

Chapter 3

Training Deep Neural Networks with Reinforcement Learning for Time Series Forecasting

Takashi Kuremoto, Takaomi Hirata, Masanao Obayashi, Shingo Mabu and Kunikazu Kobayashi

Abstract

As a kind of efficient nonlinear function approximators, artificial neural networks (ANN) have been popularly applied to time series forecasting. The training method of ANN usually utilizes error back-propagation (BP) which is a supervised learning algorithm proposed by Rumelhart et al. in 1986; meanwhile, authors proposed to improve the robustness of the ANN for unknown time series prediction using a reinforcement learning algorithm named stochastic gradient ascent (SGA) originally proposed by Kimura and Kobayashi for control problems in 1998. We also successfully use a deep belief net (DBN) stacked by multiple restricted Boltzmann machines (RBMs) to realized time series forecasting in 2012. In this chapter, a state-of-the-art time series forecasting system that combines RBMs and multilayer perceptron (MLP) and uses SGA training algorithm is introduced. Experiment results showed the high prediction precision of the novel system not only for benchmark data but also for real phenomenon time series data.

Keywords: artificial neural networks (ANN), deep learning (DL), reinforcement learning (RL), deep belief net (DBN), restricted Boltzmann machine (RBM), multilayer perceptron (MLP), stochastic gradient ascent (SGA)

1. Introduction

Artificial neural networks (ANN), which are mathematical models for function approximation, classification, pattern recognition, nonlinear control, etc., have been successfully applied in the field of time series analysis and forecasting instead of linear models such as 1970s ARIMA [1] since 1980s [2–7]. In [2], Casdagli used a radial basis function network (RBFN) which is a kind of feed-forward neural network with Gaussian hidden units to predict chaotic time series data, such as the Mackey-Glass, the Ikeda map, and the Lorenz chaos in 1989. In [3, 4], Lendasse et al. organized a time series forecasting competition for neural network prediction methods with a five-block artificial time series data named CATS since 2004. The goal of CATS competition was to predict 100 missing values of the time series data in five sets which included 980 known values and 20 successive unknown values in each set (details are in Section 3.1). There were 24 submissions to the competition, and five kinds of methods were selected by the IJCNN2004: filtering techniques including Bayesian methods, Kalman filters, and so on; recurrent neural networks

(RNNs); vector quantization; fuzzy logic; and ensemble methods. As the comment of the organizers, the different prediction precisions were reported though the similar prediction methods were used for the know-how and experience of the authors. So the development of time series forecasting by ANN is still on the way.

As a kind of classifiers or a kind of function approximators, the advances of the ANN are bought out by the nonlinear transforms to the input space. In fact, units (or neurons) with nonlinear firing functions connected to each other usually produce higher dimensional output space and various feature spaces in the networks. Additionally, as a connective system, it is not necessary to design fixed mathematical models for different nonlinear phenomena, but adjusting the weights of connections between units. So according to the report of NN3—Artificial Neural Networks and Computational Intelligence Forecasting Competition [5], there have been more than 5000 publications of time series forecasting using ANN till 2007.

To find the suitable parameters of ANN, such as weights of connections between neurons, error back-propagation (BP) algorithm [6] is generally utilized in the training process of ANN. However, due to every sample data (a pair of the input data and the output data) is used in the BP method, noise data influences the optimization of the model, and robustness of the model becomes weak for unknown input. Another problem of ANN models is how to determine the structure of the network, i.e., the number of layers and the number of neurons in each layer. To overcome these problems of BP, Kuremoto et al. [7] adopted a reinforcement learning (RL) method "stochastic gradient ascent (SGA)" [8] to adjust the connection weights of units and the particle swarm optimization (PSO) to find the optimal structure of ANN. SGA, which is proposed by Kimura and Kobayshi, improved Williams' REINFORCE [9], which uses rewards to modify the stochastic policies (likelihood). In SGA learning algorithm, the accumulated modification of policies named "eligibility trace" is used to adjust the parameters of model (see Section 2). In the case of time series forecasting, the reward of RL system can be defined as a suitable error zone to instead of the distance (error) between the output of the model and the teach data which is used in BP learning algorithm. So the sensitivity to noise data is possible to be reduced, and the robustness to the unknown data may be raised. As a deep learning method for time series forecasting, Kuremoto et al. [10] firstly applied Hinton and Salakhutdinov's deep belief net (DBN) which is a kind of stacked auto-encoder (SAE) composed by multiple restricted Boltzmann machines (RBMs) [11]. An improved DBN for time series forecasting is proposed in [12], which DBN is composed by multiple RBMs and a multilayer perceptron (MLP) [6]. The improved DBN with RBMs and MLP [6] gives its priority to the conventional DBN [5] for time series forecasting due to the continuous output unit is used; meanwhile the conventional one had a binary value unit in the output layer.

As same as the RL method, SGA adopted to MLP, RBFN, and self-organized fuzzy neural network (SOFNN) [7]; the prediction precision of DBN utilized SGA may also be raised comparing to the BP learning algorithm. Furthermore, it is available to raise the prediction precision by a hybrid model which forecasts the future data by the linear model ARIMA at first and modifying the forecasting by the predicted error given by an ANN which is trained by error time series [13, 14].

In this chapter, we concentrate to introduce the DBN which is composed by multiple RBMs and MLP and show the higher efficiency of the RL learning method SGA for the DBN [15, 16] comparing to the conventional learning method BP using the results of time series forecasting experiments. Kinds of benchmark data including artificial time series data CATS [3], natural phenomenon time series data provided by Aalto University [18], and TSDL [18] were used in the experiments.

2. The DBN model for time series forecasting

2.1 The structure of the model

The model of time series forecasting is given as the following:

$$x_{t+1} = f(x_t, x_{t-1}, \ldots, x_{t-n+1}) \tag{1}$$

Denote t = 1, 2, 3, ..., where T is the time, n is the dimensionality of the input of function $f(x)$, x_t is the time series data, and x_{t+1} is unknown data in the future as well as the output of model.

A deep belief net (DBN) composed by restricted Boltzmann machines (RBMs) and multilayer perceptron (MLP) is shown in **Figure 1**.

2.2 RBM

Restricted Boltzmann machine (RBM) is a kind of probabilistic generative neural network which composed by two layers of units: visible layer and hidden layer (see **Figure 2**).

Units of different layers connect to each other with weights $w_{ij} = w_{ji}$, where $i = 1, 2, \ldots, n$ and $j = 1, 2, \ldots, m$ are the numbers of units of visible layer and hidden layer, respectively. The outputs of units v_i, h_j are binary, i.e., 0 or 1, except for the initial value of visible units which is given by the input data. The probabilities of 1 of a visible unit and a hidden unit are according to the following:

$$p(h_j = 1|v) = \frac{1}{1 + \exp\left(-b_j - \sum_{i=1}^{n} w_{ji} v_i\right)} \tag{2}$$

$$p(v_i = 1|\text{h}) = \frac{1}{1 + \exp\left(-b_i - \sum_{j=1}^{m} w_{ij} h_j\right)} \tag{3}$$

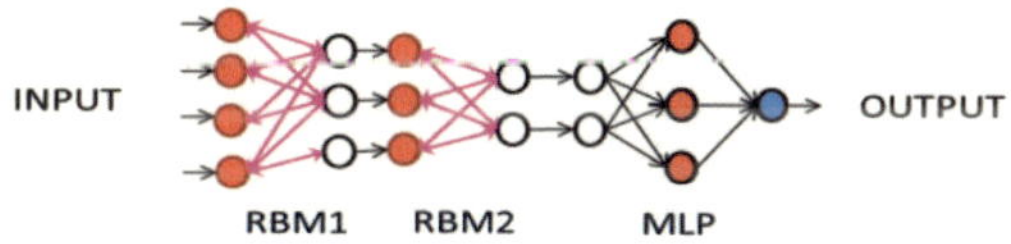

Figure 1.
The structure of DBN for time series forecasting.

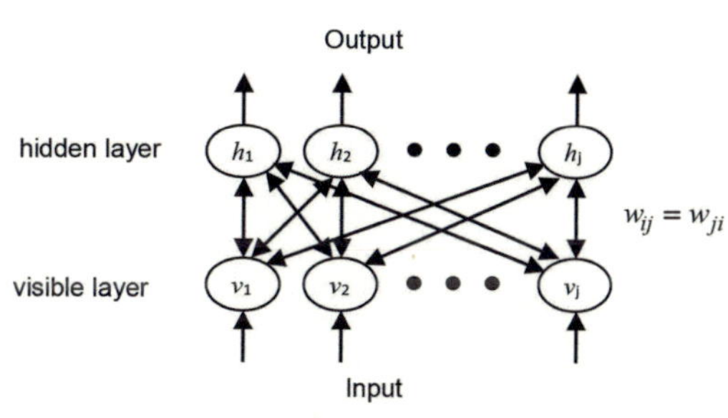

Figure 2.
The structure of RBM.

Here b_i, b_j are the biases of units. The learning rules of RBM are given as follows:

$$\Delta w_{ij} = \varepsilon\left(< v_i h_j >_{\mathbf{data}} - < v_i h_j >_{\mathbf{model}} \right) \tag{4}$$

$$\Delta b_i = \varepsilon\left(< v_i > - < \tilde{v}_i > \right) \tag{5}$$

$$\Delta b_j = \varepsilon\left(< h_j > - < \tilde{h}_j > \right) \tag{6}$$

where $0 < \varepsilon < 1$ is a learning rate, $p_{ij} = < v_i h_j >_{\mathbf{data}}$, $p'_{ij} < v_i h_j >_{\mathbf{model}}$ and $< v_i >$, $< h_j >$ indicate the expectations of the first Gibbs sampling ($k = 0$), and $< \tilde{v}_i >$, $< \tilde{h}_j >$ the kth Gibbs sampling, and it works when $k = 1$.

2.3 MLP

Multilayer perceptron (MLP) is the most popular neural network which is generally composed by three layers of units: input layer, hidden layer, and output layer (see **Figure 3**).

The output of the unit $y = f(z)$ and unit $z_k = f(\mathrm{x})$ is given as the following logistic sigmoid functions:

$$f(y) = \frac{1}{1 + \exp\left(-\sum_{j=1}^{K+1} w_j z_j \right)} \tag{7}$$

$$f(z_j) = \frac{1}{1 + \exp\left(-\sum_{i=1}^{n+1} v_{ji} x_i \right)} \tag{8}$$

Here n is the dimensionality of the input, K is the number of hidden units, and $x_{n+1} = 1.0, z_{K+1} = 1.0$ are the support units of biases $v_{j(n+1)}$, w_{K+1}.

The learning rules of MLP using error back-propagation (BP) method [5] are given as follows:

$$\Delta w_j = -\varepsilon(y - \tilde{y})y(1 - y)z_j \tag{9}$$

$$\Delta v_{ji} = -\varepsilon(y - \tilde{y})y(1 - y)w_j z_j (1 - z_j) x_i \tag{10}$$

where $0 < \varepsilon < 1$ is a learning rate and $\tilde{y}$ is the teacher signal.

The learning algorithm of MLP using BP is as follows:

Step 1. Observe an input $\mathrm{x}_t = (x_t, x_{t-1}, \ldots, x_{t-n+1})$;

Step 2. Predict a future data $y_t = x_{t+1}$ according to Eqs. (7) and (8).

Step 3. Calculate the modification of connection weights, Δw_j, Δv_{ji} according to Eqs. (9) and (10).

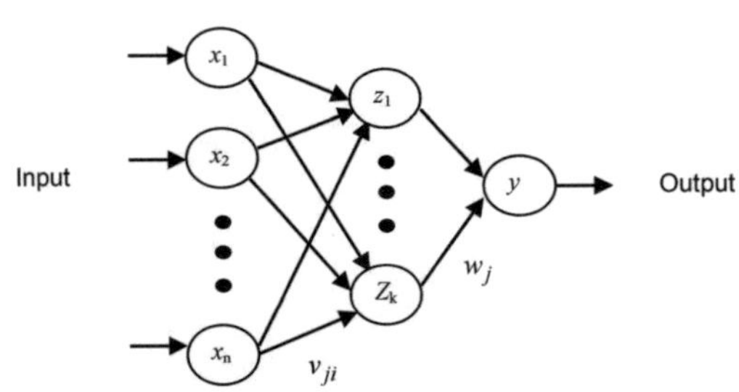

Figure 3.
The structure of MLP.

Step 4. Modify the connections,

$$w_j \leftarrow w_j + \Delta w_j; v_{ji} \leftarrow v_{jj} + \Delta v_{ji};$$

Step 5. For the next time step $t + 1$, return to step 1.

2.4 The training method of DBN

As same as the training process proposed in [10], the training process of DBN is performed by two steps. The first one, pretraining, utilizes the learning rules of RBM, i.e., Eqs. (4–6), for each RBM independently. The second step is a fine-tuning process using the pretrained parameters of RBMs and BP algorithm. These processes are shown in **Figure 4** and Eqs. (11)–(13).

$$\Delta w_{ji}^{L} = -\varepsilon\left(\sum_i \frac{\partial E}{\partial w_{ji}^{L+1}} w_{ji}^{L+1}\right)\left(1 - h_j^L\right)v_i^L \tag{11}$$

$$\Delta b_j^{L} = -\varepsilon\left(\sum_i \frac{\partial E}{\partial w_{ji}^{L+1}} w_{ji}^{L+1}\right)\left(1 - h_j^L\right) \tag{12}$$

$$E = \frac{1}{2}\sum_{t=1}^{T}\left(y_t - \tilde{y}_t\right) \tag{13}$$

In the case of reinforcement learning (RL), the output is decided by a probability distribution, e.g., the Gaussian distribution $y \sim \pi(\mu, \sigma^2)$. So the output units are the mean μ and the variance σ instead of one unit y.

$$\mu = \sum_j w_{\mu j} z_j \tag{14}$$

$$\sigma = \frac{1}{1 + \exp\left(-\sum_j w_{\sigma j} z_j\right)} \tag{15}$$

$$\pi\left(\mu, \sigma^2\right) = \frac{1}{\sqrt{2\pi}\sigma} \exp\left(-\frac{(y-\mu)^2}{2\sigma^2}\right) \tag{16}$$

The learning algorithm of stochastic gradient ascent (SGA) [7] is as follows.

Step 1. Observe an input $\mathrm{x}_t = (x_t, x_{t-1}, \ldots, x_{t-n+1})$.

Step 2. Predict a future data $y_t = x_{t+1}$ according to a probability $y_t \sim \pi(\mathrm{x}_t, \mathrm{w})$ with ANN models which are constructed by parameters $\mathrm{w}\left(w_{\mu j}, w_{\sigma j}, w_{ij}, v_{ji}\right)$.

Step 3. Receive a scalar reward/punishment r_t by calculating the prediction error:

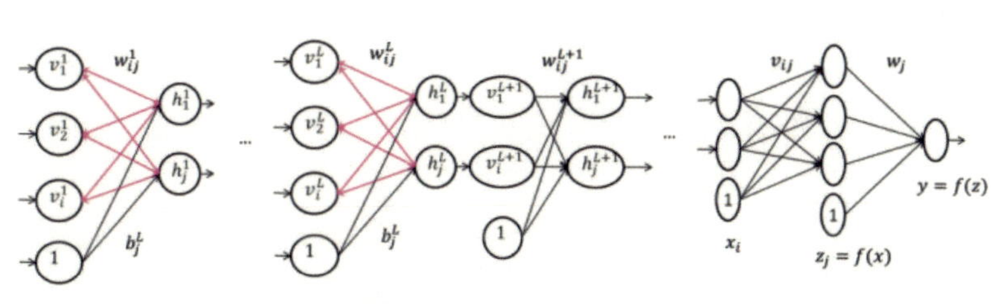

Figure 4.
The training of DBN by BP method.

$$r_t = \begin{cases} 1 & if\, (y_t - \tilde{y}_t)^2 \leq \zeta \\ -1 & else \end{cases} \tag{17}$$

where ζ is an evaluation constant greater than or equal to zero.

Step 4. Calculate characteristic eligibility $e_i(t)$ and eligibility trace $\overline{D}_i(t)$:

$$e_i(t) = \frac{\partial}{\partial w_i} \ln \{\pi(\mathbf{x}_t, \mathbf{w})\} \tag{18}$$

$$\overline{D}_i(t) = e_i(t) + \gamma \overline{D}_i(t-1) \tag{19}$$

where $0 \leq \gamma < 1$ is a discount factor and w_i denotes ith internal variable of DBN.

Step 5. Calculate the modification $\Delta w_i(t)$:

$$\Delta w_i(t) = (r_t - b)\overline{D}_i(t) \tag{20}$$

where $b \geq 0$ denotes the reinforcement baseline (it can be set as zero).

Step 6. Improve the policy Eq. (16) by renewing its internal variable w_i by Eq. (21):

$$w_i \leftarrow w_i + \varepsilon \Delta w_i \tag{21}$$

where $0 \leq \varepsilon \leq 1$ is a learning rate.

Step 7. For the next time step $t+1$, return to step 1.

Characteristic eligibility $e_i(t)$, shown in Eq. (18), means that the change of the policy function concerns with the change of system internal variable vector. In fact, the algorithm combines reward/punishment to modify the stochastic policy with its internal variable renewing by Step 4 and Step 5.

The calculation of $e_{w_{\mu j}}(t)$, $e_{w_{\sigma j}}(t)$, $e_{v_{ij}}(t)$ in MLP part of DBN is induced as follows;

$$e_{w_{\mu j}}(t) = \frac{y_t - \mu_t}{\sigma_t^2} z_j(t) \tag{22}$$

$$e_{w_{\sigma j}}(t) = \frac{(y_t - \mu_t)^2 - \sigma^2}{\sigma_t^2} (1 - \sigma_t) z_j(t) \tag{23}$$

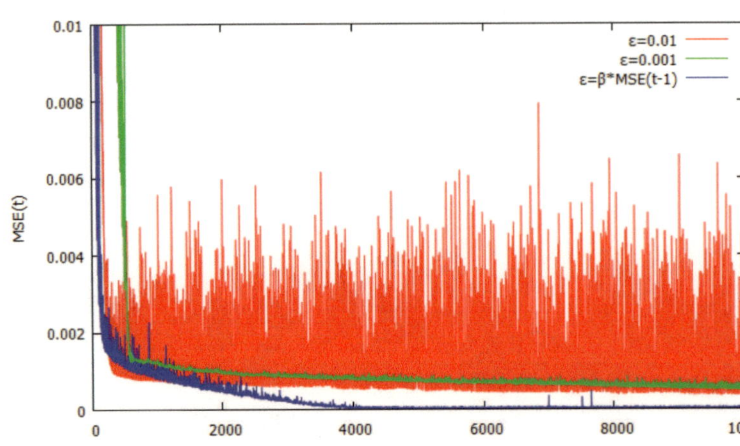

Figure 5.
The learning errors given by different learning rates.

$$e_{v_{ij}}(t) = \left(e_{w_{\mu j}}(t)w_{\mu j} + e_{w_{\sigma j}}(t)w_{\sigma j}\right)(1 - z_j(t))x_i(t) \tag{24}$$

The $e_i(t)$ of the RBM of Lth layer in the case of the DBN is given as follows:

$$e^L_{w_{ij}}(t) = \left(\sum_j e^{L+1}_{w_{ij}}(t)w^{L+1}_{w_{ij}}\right)\left(1 - h^L_j\right)v^L_i \tag{25}$$

$$e^L_{b_j}(t) = \left(\sum_k e^{L+1}_{w_{ij}}(t)w^{L+1}_{w_{ij}}\right)\left(1 - h^L_j\right) \tag{26}$$

The learning rate ε in Eq. (21) affects the learning performance of fine-tuning of DBN. Different values to result different training error (mean squared error (MSE)) as shown in **Figure 5**. An adaptive learning rate as a linear function of learning error is proposed as in Eq. (27):

$$\varepsilon = \beta \text{MSE}(t - 1) \tag{27}$$

where is $0 \leq \beta$ a constant.

2.5 Optimization of meta-parameters

The number of RBM that constitute the DBN and the number of neurons of each layer affects prediction performance seriously. In [9], particle swarm optimization (PSO) method is used to decide the structure of DBN, and in [13] it is suggested that random search method [16] is more efficient. In the experiment of time series forecasting by DBN and SGA shown in this chapter, these meta-parameters were decided by the random search, and the exploration limits are shown as the following.

- The number of RBMs: [0–3]
- The number of units in each layer of DBN: [2–20]
- Learning rate of each RBM in Eqs. (4)–(6): [10^{-5}–10^{-1}]
- Fixed learning rate of SGA in Eq. (21): [10^{-5}–10^{-1}]
- Discount factor in Eq. (19): [10^{-5}–10^{-1}]
- Coefficient in Eq. (27) [0.5–2.0]

The optimization algorithm of these meta-parameters by the random search method is as follows:

Step 1. Set random values of meta-parameters beyond the exploration limitations.

Step 2. Predict a future data $y_t \approx x_{t+1}$ by MLP or DBN using the current weighted connections.

Step 3. If the error between y_t, x_{t+1} is reduced enough, store the values of meta-parameters,
or else if the error is not changed,
stop the exploration,
else return to step 1.

3. The experiments and results

3.1 CATS benchmark time series data

CATS time series data is the artificial benchmark data for forecasting competition with ANN methods [3, 4].This artificial time series is given with 5000 data, among which 100 are missed (hidden by competition the organizers). The missed data exist in five blocks:

- Elements 981 to 1000
- Elements 1981 to 2000
- Elements 2981 to 3000
- Elements 3981 to 4000
- Elements 4981 to 5000

The mean square error E_1 is used as the prediction precision in the competition, and it is computed by the 100 missing data and their predicted values as the following:

$$E_1 = \left\{ \sum_{t=981}^{1000} (y_t - \bar{y}_t)^2 + \sum_{t=1981}^{2000} (y_t - \bar{y}_t)^2 + \sum_{t=2981}^{3000} (y_t - \bar{y}_t)^2 + \sum_{t=3981}^{4000} (y_t - \bar{y}_t)^2 + \sum_{t=4981}^{5000} (y_t - \bar{y}_t)^2 \right\} / 100 \quad (28)$$

where $\bar{y}_t$ is the long-term prediction result of the missed data. The CATS time series data is shown in **Figure 6**.

The prediction results of different blocks of CATS data are shown in **Figure 7**. Comparing to the conventional learning method of DBN, i.e., using Hinton's RBM unsupervised learning method [6, 8] and back-propagation (BP), the proposed method which used the reinforcement learning method SGA instead of BP showed its superiority in the sense of the average prediction precision E_1 (see **Figure 7f**).

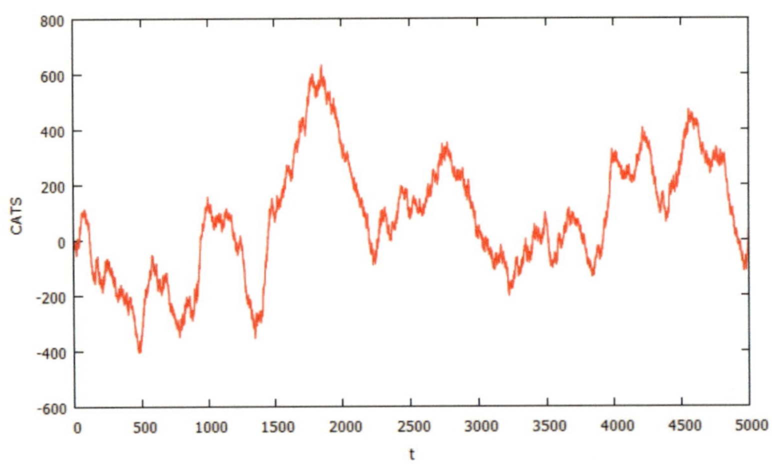

Figure 6.
CATS benchmark data.

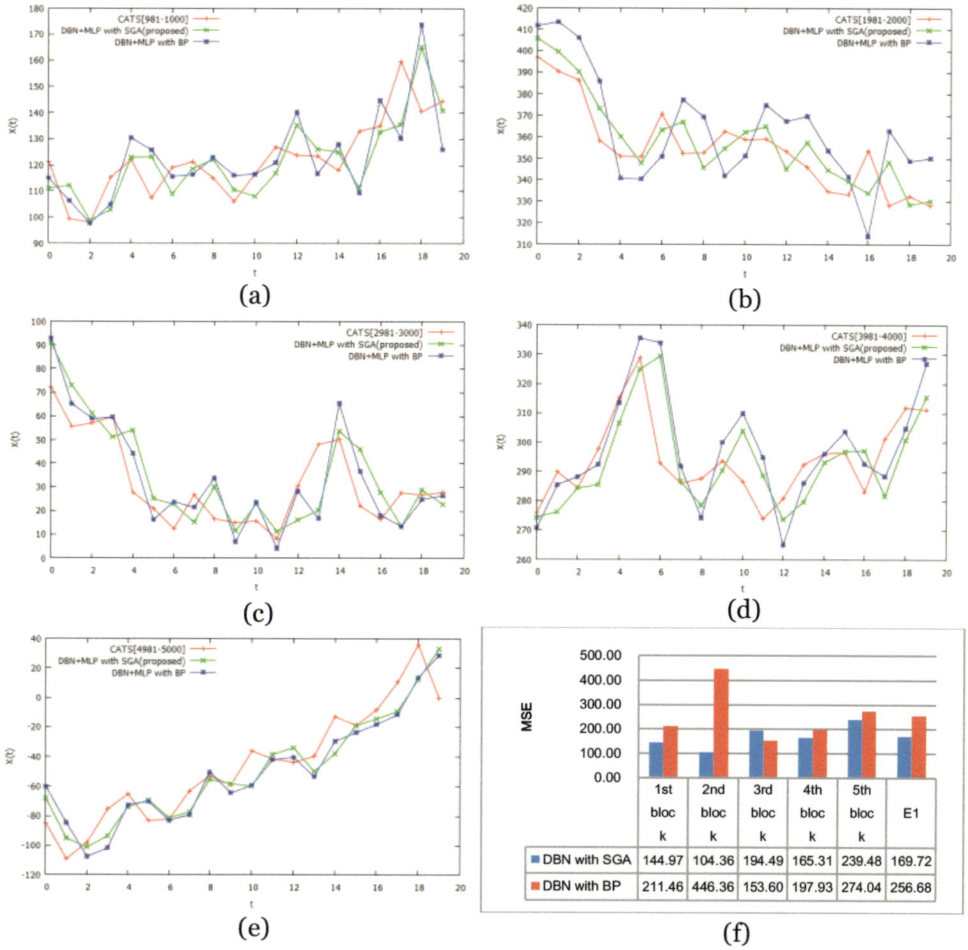

Figure 7.
The prediction results of different methods for CATS data: (a) block 1; (b) block 2; (c) block 3; (d) block 4; (e) block 5; and (f) results of the long-term forecasting.

In addition, the proposed method, DBN with SGA, yielded the highest prediction (E1 measurement) comparing to all previous studies such as MLP with BP, the best prediction of CATS competition IJCNN'04 [4], the conventional DBNs with BP [9, 11], and hybrid models [13]. The details are shown in **Table 1**.

The meta-parameters obtained by random search method are shown in **Table 2**. And we found that the MSE of learning, i.e., given by one-ahead prediction results, showed that the proposed method has worse convergence compared to the conventional BP training. In **Figure 8**, the case of the first block learning MSE of two methods is shown. The convergence of MSE given by BP converged in a long training process and SGA gave unstable MSE of prediction. However, as the basic consideration of a sparse model, the better results of long-term prediction of the proposed method may successfully avoid the over-fitting problem which is caused by the model that is built too strictly by the training sample and loses its robustness for unknown data.

3.2 Real time series data

Three types of natural phenomenon time series data provided by Aalto University [17] were used in the one-ahead forecasting experiments of real time series data.

Method	E_1
DBN(SGA) [18]	170
DBN(BP) + ARIMA [14]	244
DBN [11] (BP)	257
Kalman Smoother (the best of IJCNN '04) [4]	408
DBN [9] (2 RBMs)	1215
MLP [9]	1245
A hierarchical Bayesian learning (the worst of IJCNN '04) [4]	1247
ARIMA [1]	1715
ARIMA+MLP(BP) [12]	2153
ARIMA+DBN(BP) [14]	2266

Table 1.
The long-term forecasting error comparison of different methods using CATS data.

	DBN with SGA	DBN with BP
The number of RBMs	3	1
Learning rate of RBM	0.048-0.055-0.026	0.042
Structure of DBN (the number of units and layers)	14-14-18-19-18-2	5-11-2-1
Learning rate of SGA or BP	0.090	0.090
Discount factor γ	0.082	—
Coefficient β	1.320	—

Table 2.
Meta-parameters of DBN used for the CATS data (block 1).

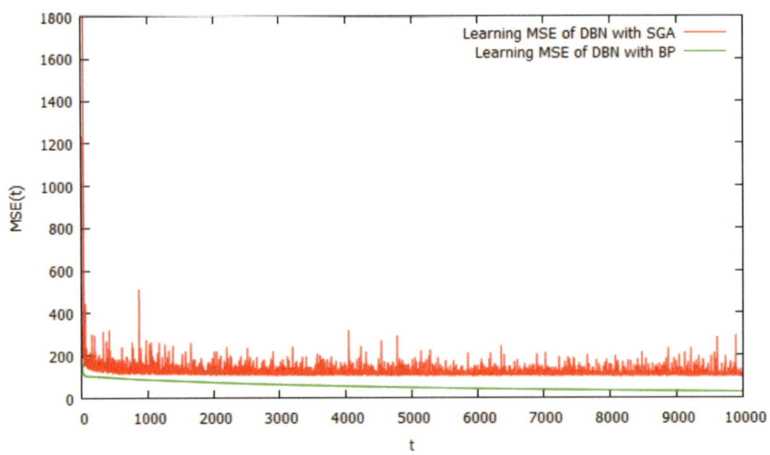

Figure 8.
Change of the learning error during fine-tuning (CATS data [1–980]).

- CO_2: Atmospheric CO_2 from continuous air samples weekly averages atmospheric CO_2 concentration derived from continuous air samples, Hawaii, 2225 data

- Sea level pressures: Monthly values of the Darwin sea level pressure series, A.D. 1882–1998, 1300 data

- Sunspot number: Monthly averages of sunspot numbers from A.D. 1749 to the present 3078 values

The prediction results of these three datasets are shown in **Figure 9**. Short-term prediction error is shown in **Table 3**. DBN with the SGA learning method showed its priority in all cases.

The efficiency of random search to find the optimal meta-parameters, i.e., the structure of RBM and MLP, learning rates, discount factor, etc. which are explained in Section 2.5 is shown in **Figure 10** in the case of DBN with SGA learning algorithm. The random search results are shown in **Table 4**.

We also used seven types of natural phenomenon time series data of TSDL [18]. The data to be predicted was chosen based on [19] which are named as Lynx, Sunspots, River flow, Vehicles, RGNP, Wine, and Airline. The short-term (one-ahead) prediction results are shown in **Figure 11** and **Table 5**.

From **Table 5**, it can be confirmed that SGA showed its priority to BP except the cases of Vehicles and Wine. From **Table 6**, an interesting result of random search for meta-parameter showed that the structures of DBN for different datasets were different, not only the number of units on each layer but also the number of RBMs. In the case of SGA learning method, the number of layer for Sunspots, River flow, and Wine were more than DBN using BP learning.

4. Discussions

The experiment results showed the DBN composed by multiple RBMs and MLP is the state-of-the-art predictor comparing to all conventional methods in the case of CATS data. Furthermore, the training method for DBN may be more efficient by the RL method SGA for real time series data than using the conventional BP algorithm. Here let us glance back at the development of this useful deep learning method.

- Why the DBN composed by multiple RBMs and MLP [11, 13] is better than the DBN with multiple RBMs only [9]?

The output of the last RBM of DBN, a hidden unit of the last RBM in DBN, has a binary value during pretraining process. So the weights of connections between the unit and units of the visible layer of the last RBM are affected and with lower complexity than using multiple units with continuous values, i.e., MLP, or so-called full connections in deep learning architecture.

- How are RL methods active at ANN training?

In 1992, Williams proposed to adopt a RL method named REINFORCE to modify artificial neural networks [8]. In 2008, Kuremoto et al. showed the RL method SGA is more efficient than the conventional BP method in the case of time series forecasting [6]. Recently, researchers in DeepMind Ltd. adopted RL into deep neural networks and resulted a famous game software AlphaGo [20–23].

- Why SGA is more efficient than BP?

Generally, the training process for ANN by BP uses mean square error as loss function. So every sample data affects the learning process and results including noise data. Meanwhile, SGA uses reward which may be an error zone to modify the

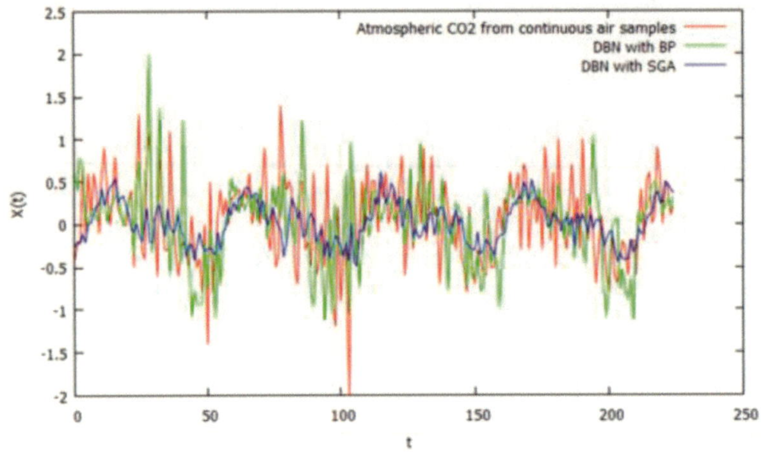

(a)

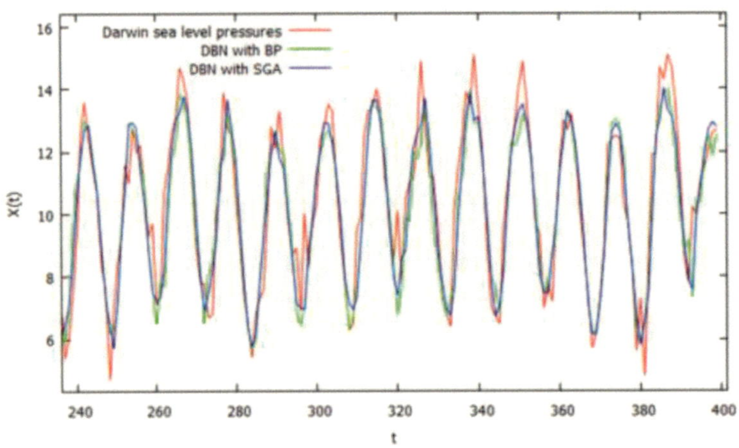

(b)

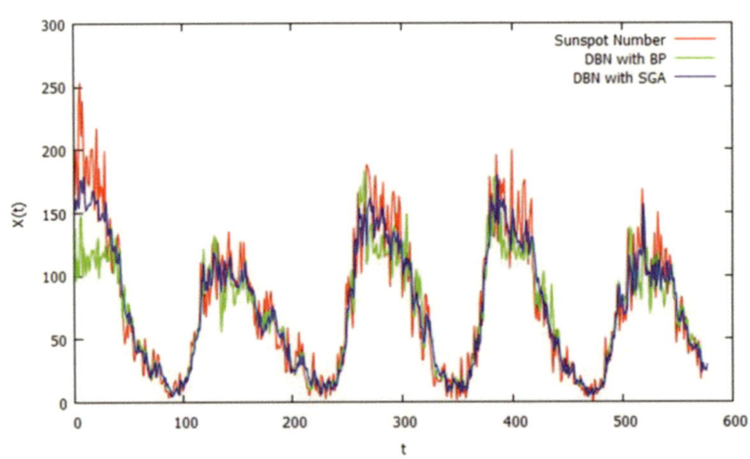

(c)

Figure 9.
Prediction results by DBN with BP and SGA. (a) Prediction result of CO_2 *data. (b) Prediction result of Sea level pressure data. (c) Prediction result of Sun spot number data.*

Data	DBN with BP	DBN with SGA
CO2	0.2671	0.2047
Sea level pressure	0.9902	0.9003
Sun spot number	733.51	364.05

Table 3.
Prediction MSE of real time series data [17].

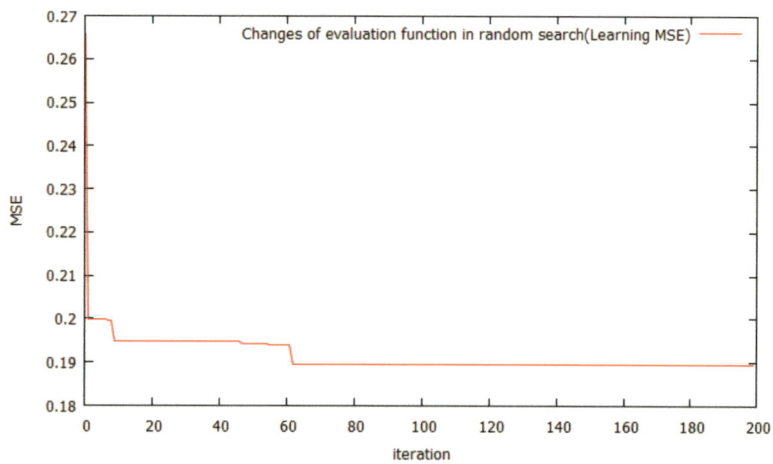

Figure 10.
Changes of learning error by random search for DBN with SGA.

Data series	Total data	Testing data	DBN with BP (the number of units)	DBN with SGA (the number of units)
CO_2	2225	225	15-17-17-1	20-18-7-2
Sea level pressure	1400	400	16-18-18-1	16-20-8-7-2
Sun spot number	3078	578	20-20-17-18-1	19-19-20-10-2

Table 4.
Meta-parameters of DBN used for real time series forecasting.

parameters of model. So it has higher robustness for the noisy data and unknown data for real problems.

5. Conclusions

A deep belief net (DBN) composed by multiple restricted Boltzmann machines (RBMs) and multilayer perceptron (MLP) for time series forecasting were introduced in this chapter. The training method of DBN is also discussed as well as a reinforcement learning (RL) method; stochastic gradient ascent (SGA) showed its priority to the conventional error back-propagation (BP) learning method. The robustness of SGA comes from the utilization of relaxed prediction error during the

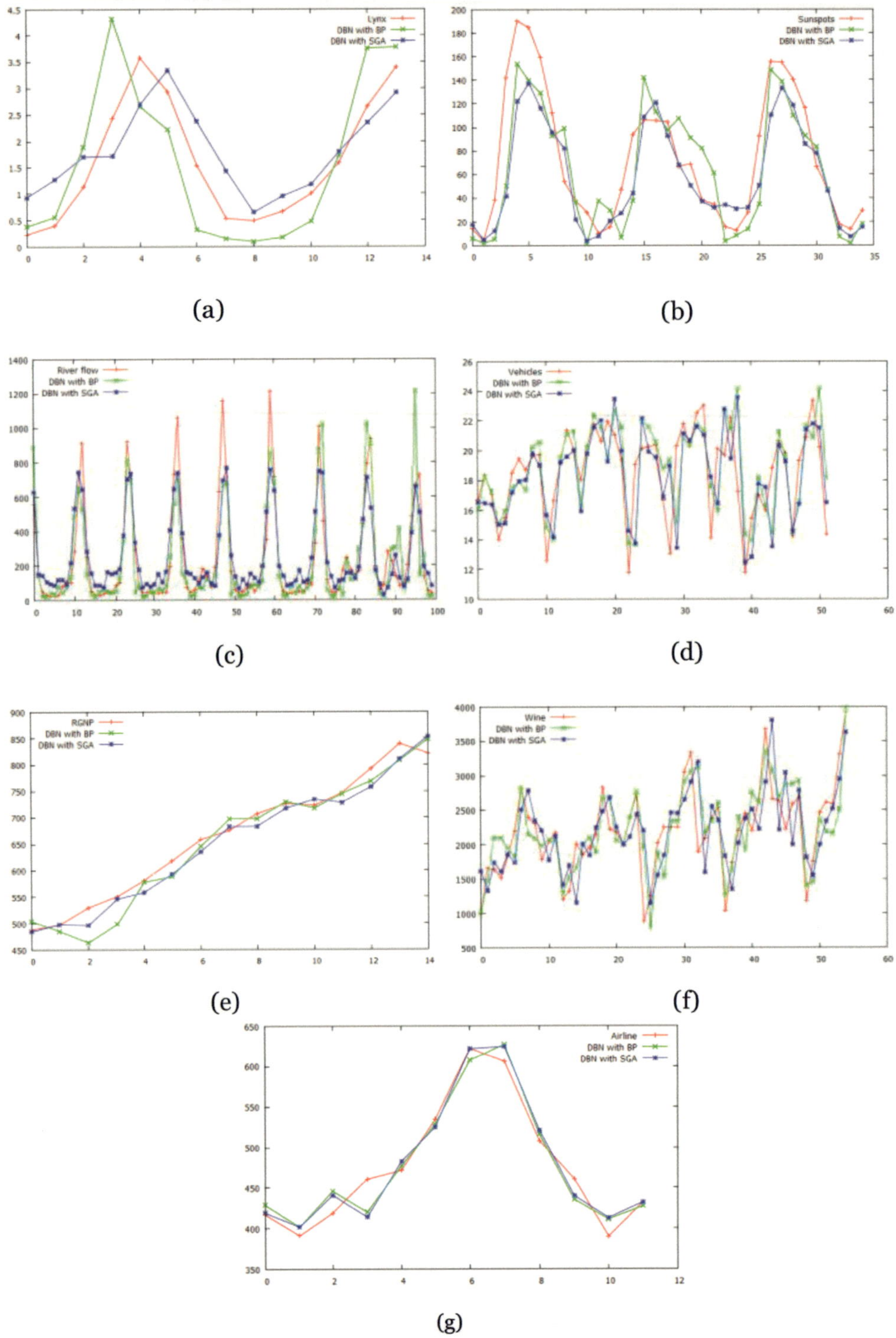

Figure 11.
Prediction results of natural phenomenon time series data of TSDL. (a) Prediction result of Lynx; (b) prediction result of sunspots; (c) prediction result of river flow; (d) prediction result of vehicles; (e) prediction result of RGNP; (f) prediction result of wine; and (g) prediction result of airline.

learning process, i.e., different from the BP method which adopts all errors of every sample to modify the model. Additionally, the optimization of the structure of DBN was realized by random search method. Time series forecasting experiments used

Data	DBN with BP	DBN with SGA
Lynx	0.6547	**0.3593**
Sunspots	999.54	**904.35**
River flow	24262.24	**16980.46**
Vehicles	**6.0670**	6.1919
RGNP	771.79	**469.72**
Wine	**138743.80**	224432.02
Airline	380.60	**375.25**

Table 5.
Prediction MSE of time series data of TSDL.

Series	Total data	Testing data	DBN with BP	DBN with SGA
Lynx	114	14	19-16-1	7-14-2
Sunspots	288	35	20-18-11-1	10-12-12-17-2
River flow	600	100	20-17-18-1	19-20-5-18-5-2
Vehicles	252	52	20-13-20-1	20-11-5-2
RGNP	85	15	18-20-1	19-15-2
Wine	187	55	16-15-12-1	18-12-13-11-2
Airline	144	12	15-4-1	13-7-2

Table 6.
Size of time series data and structure of prediction network.

benchmark CATS data, and real time series datasets showed the effectiveness of the DBN. As for the future work, there are still some problems that need to be solved such as how to design the variable learning rate and reward which influence the learning performance strongly and how to prevent the explosion of characteristic eligibility trace in SGA.

Author details

Takashi Kuremoto[1]*, Takaomi Hirata[1], Masanao Obayashi[1], Shingo Mabu[1] and Kunikazu Kobayashi[2]

1 Yamaguchi University, Ube, Japan

2 Aichi Prefectural University, Nagakute, Japan

*Address all correspondence to: wu@yamaguchi-u.ac.jp

References

[1] Box GEP, Pierce DA. Distribution of residual autocorrelations in autoregressive-integrated moving average time series models. Journal of the American Statistical Association. 1970;**65**(332):1509-1526

[2] Casdagli M. Nonlinear prediction of chaotic time series. Physica D. 1989;**35**: 335-356

[3] Lendasse A, Oja E, Simula O, Verleysen M. Time series prediction competition: The CATS benchmark. In: Proceedings of International Joint Conference on Neural Networks (IJCNN'04); 2004. pp. 1615-1620

[4] Lendasse A, Oja E, Simula O, Verleysen M. Time series prediction competition: The CATS benchmark. Neurocomputing. 2007;**70**:2325-2329

[5] NN3. http://www.neural-foreca sting-competition.com/NN3/index.htm

[6] Rumelhart DE, Hinton GE, Williams RJ. Learning representation by back-propagating errors. Nature. 1986; **232**(9):533-536

[7] Kuremoto T, Obayashi M, Kobayashi M. Neural forecasting systems, Chapter 1. In: Weber C, Elshaw M, Mayer NM, editors. Reinforcement Learning, Theory and Applications. Rijeka, Croatia: InTech; 2008. pp. 1-20

[8] Kimura H, Kobayashi S. Reinforcement learning for continuous action using stochastic gradient ascent. In: Proceedings of 5th Intelligent Autonomous Systems (IAS-5); 1998. pp. 288-295

[9] Williams RJ. Simple statistical gradient following algorithms for connectionist reinforcement learning. Machine Learning. 1992;**8**:229-256

[10] Kuremoto T, Kimura S, Kobayashi K, Obayashi M. Time series forecasting using a deep belief network with restricted Boltzmann machines. Neurocomputing. Aug. 2014;**137**(5): 47-56

[11] Hinton GE, Salakhutdinov RR. Reducing the dimensionality of data with neural networks. Science. 2006; **313**:504-507

[12] Kuremoto T, Hirata T, Obayashi M, Mabu S, Kobayashi K. Forecast chaotic time series data by DBNs. In: Proceedings of the 7th International Congress on Image and Signal Processing (CISP 2014); Oct. 2014. pp. 1304-1309

[13] Zhang GP. Time series forecasting using a hybrid ARIMA and neural network model. Neurocomputing. 2003; **50**:159-175

[14] Hirata T, Kuremoto T, Obayashi M, Mabu S. Time series prediction using DBN and ARIMA. In: Proceedings of International Conference on Computer Application Technologies (CCATS 2015). Matsue, Japan; Sep. 2015. pp. 24-29

[15] Hirata T, Kuremoto T, Obayashi M, Mabu S, Kobayashi K. Deep belief network using reinforcement learning and its applications to time series forecasting. In: Proceedings of International Conference on Neural Information Processing, (ICONIP'16), Lecture Notes in Computer Science (LNCS). Heidelberg, Germany: Springer. Vol. 9949. Kyoto, Japan; Oct. 18–21, 2016. pp. 30-37

[16] Hirata T, Kuremoto T, Obayashi M, Mabu S, Kobayashi K. Forecasting real time series data using deep belief net and reinforcement learning. Journal of Robotics, Network and Artificial Life.

2018;**4**(4):260-264. DOI: 10.2991/jrnal.2018.4.4.1

[17] Bergstra J, Bengio Y. Random search for hyper-parameter optimization. Journal of Machine Learning Research. 2012;**13**:281-305

[18] Aalto University Applications of Machine Learning Group Datasets. Available online at url: <http://research.ics.aalto.fi/eiml/datasets.shtml> (01-01-17)

[19] Hyndman RJ. Time Series Data Library (TSDL). 2013. Available online at url: ⟨http://robjhyndman.com/TSDL/⟩ (01-01-13)

[20] Adhikari R. A neural network based linear ensemble framework for time series forecasting. Neurocomputing. 2015;**157**:231-242

[21] Mnih V et al. Human-level control through deep reinforcement learning. Nature. 2015;**518**:529-533

[22] Sivler D et al. Mastering the game of go with deep neural networks and tree search. Nature. 2017;**529**:484-489

[23] Sivler D et al. Mastering the game of go without human knowledge. Nature. 2017;**550**:354-359

Chapter 4

CNN Approaches for Time Series Classification

Lamyaa Sadouk

Abstract

Time series classification is an important field in time series data-mining which have covered broad applications so far. Although it has attracted great interests during last decades, it remains a challenging task and falls short of efficiency due to the nature of its data: high dimensionality, large in data size and updating continuously. With the advent of deep learning, new methods have been developed, especially Convolutional Neural Network (CNN) models. In this paper, we present a review of our time series CNN approaches including: (i) a data-level approach based on encoding time series into frequency-domain signals via the Stockwell transform, (ii) an algorithm-level approach based on an adaptive convolutional layer filter that suits the time series in hand, and (iii) another algorithm-level approach adapted to time series classification tasks with limited annotated data, which is a global, fast and light-weight framework based on a transfer learning technique with a source learning task similar or different but related to the target learning task. These approaches are implemented on identifying human activities including normal movements of typical subjects and disorder-related movements such as stereotypical motor movements of autistic subjects. Experimental results show that our approaches improve performance of time series classification.

Keywords: time series, classification, convolutional neural networks, transfer learning

1. Introduction

Time series is a series of data points which are collected by recordinga set of observations chronologically. Examples of time series include speech, human activities, electrocardiogram (ECG), etc. Recently, time series classification has attracted great interests and initiated various researches. However, the nature of time series data, including the large size of data, the high dimensionality and the continuously updating scheme of time series, makes time series classification a more challenging task.

Time series classification is widely applied in different fields such as in astronomy [1] to classify the brightness of a target star, in medical science to diagnose cardiac disorders [2] or to recognize human activities [3, 4], and in computer science for speech recognition [5, 6]. To handle time series classification, several techniques were proposed, which can be aggregated into three categories: model based, distance based and feature based.

The first category of time series classification approaches consists of building a model for each class by fitting its parameters to that class. Examples of such approaches are the autoregressive (AR) model [7] and the hidden Markov model (HMM) [8] which are limited to stationary and symbolic non-stationary time series respectively.

The second category relies on developing distance functions to measure the similarity (or dissimilarity) between two time series and on selecting a good classifier, such as dynamic time warping (DTW) distance [9, 10]. But these approaches are computationally expensive.

The third category consists of extracting meaningful features from the time series. Examples of such approaches include the discrete Fourier transform (DFT) [11], the Short-time Fourier transform (STFT) [12], the discrete wavelet transform (DWT), principal component analysis (PCA), singular value decomposition (SVD), sparse coding [13], and shapelets [14].

Meanwhile, automatic feature-based approaches using deep learning models rely have been successfully applied to time series classification, classification problems, especially convolutional neural networks (CNNs) which are regarded as the most successful and commonly used deep learning model. In [5, 6], authors address the problematic of speech recognition whereby speech signals have similar patterns within different frequency band locations which convey a different meaning. A solution to this problem is to employ a limited weight sharing CNN [6] where weight sharing is limited only to local filters which are close to each other and which are pooled together in the subsampling layer. Another approach based on tiled CNN architecture with a pre-training stage (an unsupervised learning algorithm named topographic ICA) was proposed by [15], which showed its superiority over traditional CNN on small time series datasets. A tiled CNN [16] is a CNN which unties weights locally and uses a regular "tiled" pattern of tied weights that requires only that hidden units *k* steps away from each other to have tied weights. Another relevant CNN architecture for time series classification named multi-scale convolutional neural network (MCNN) [17] was introduced where each of the three transformed versions of the input (which will be discussed in Section 3.1) is fed into a branch i.e., a set of consecutive convolutional and pooling layers, resulting in three outputs which are concatenated and further fed into more convolutional and pooling layers, fully connected layers and a softmax layer to generate the final output. Training all parameters is done jointly using back-propagation. Another attempt to enhance time series classification was proposed in [18], which employs the same idea of multiple branches within the CNN architecture, except that the input is not a different transformed version of the time series signal fed into each branch, but rather a duplicate of the same time series signal fed into all the branches (three branches). However, different convolutional filter sizes are applied per branch in order to capture the multi-scale characteristics of the time series. Two other CNN proposals to time series classification were suggested in [19], namely fully convolutional networks (FCN) without subsampling layers, and residual networks (ResNet). FCNs [20] are defined as networks which have convolutional layers only and no fully-connected layers, whereas ResNet [21] is a type of specialized neural network that solves the "vanishing gradient" problem when having many layers within the network, by using residual blocks which take advantage of residual mapping to preserve inputs. By adding batch normalization layers into FCN and ResNet, and by replacing the fully connected layers with a global pooling layer in the FCN, these two deep learning models seem to yield comparable or better results than MCNN [17]. An ensemble method of deep learning networks named LSTM-FCN is proposed in [22] is proposed and consists of feeding the same time series input into two branches: an FCN and Long Short Term Recurrent Neural Network (LSTM) block [23], producing two outputs which are concatenated and then passed onto a softmax classification layer. Another attempt to helps the CNN converge faster and better to the minima was made by Guennec et al. [24] who propose to perform data-augmentation techniques (further described in Section 3.1) and pre-train each layer in an unsupervised manner (using an auto-encoder) using unlabeled training time series from different datasets. For multivariate time series, only few research papers based on CNNs were published (such as [3, 4, 25, 26]). Zheng et al. [25] proposed a multi-channels deep convolution neural network (MC-DCNN), each branch of which takes a single dimension of the multivariate time series as input and learns features individually. Then the MC-DCNN model combines the learnt features of each branch and feeds them into a fully connected layer to perform classification. And, to further improve the performance, authors also suggested to pre-train the MC-DCNN first by applying an unsupervised initialization via the convolutional auto-encoder method. Meanwhile, a different CNN architecture for multivariate time series classification was introduced in [3, 4, 26], which treats the 3-, 12-, and 9-variate time series inputs (in [3, 4, 26] respectively) as a 3-, 12-, and 9-channel inputs and convolves them as a whole instead of convolving each channel of the input separately as performed in [25]. Authors of this architecture argue that, by separating multivariate time series into univariate ones just as in [25], the interrelationship between different univariate time series may be lost and thus will not be mined/extracted.

In this paper, we aim at presenting a review on our CNN approaches for time series classification. Our review discusses our CNN contributions at the data-level and at the algorithm-level. Our paper is organized as follows. In Section 2, some preliminaries about time series are introduced. In Section 3 reviews existent data-level techniques are presented, our data-level technique is reviewed, and experiments as well as results of our technique are laid out. Section 4 describes our algorithm-level approaches for time series classification, with experiments conducted and results analyzed. Section 4.3.3 concludes our paper with future perspectives.

2. Preliminaries/nature of time series data

Univariate and multivariate time series data. Time series inputs can be categorized into: (i) Univariate Time series which have only a single variable observed at each time and thus resulting in one channel per time series input, and (ii) Multivariate Time series which have two or more variables observed at each time, ending up with multiple channels per time series input. Most time series analysis methods focus on univariate data as it is the simplest to work with. Multivariate time series analysis considers simultaneously multiple time series, which, in general is much more complicated than univariate time series analysis as it is harder to model and often many of the classical methods do not perform well.

Raw data or extracted signals. A raw time series is a series of data points indexed in time order i.e., a sequence of discrete-time data taken at successive equally spaced points in time. In time series classification tasks, some authors choose to evaluate the performance of their approaches using raw time series data taken from a specific field/domain while some others prefer to use public datasets in which the raw time series is already segmented and converted into a set of fixed-length signals. Indeed, several research papers using CNNs [17–19, 22, 24, 26, 27] build their experimental studies on the UCR time series classification archive [28] which consists of extracted short signals. Nonetheless, this benchmark is composed of relatively small datasets (with a small number of instances), which makes the CNN less efficient knowing that CNNs require large training sets for training. Furthermore, in most of the cases, fixed-length signals cannot be further encoded into new representations (which are discussed in Section 3.1), as opposed to raw time series. These issues have led authors of [3–6, 15, 25, 29, 30] to use raw time series data instead.

3. Data-level approach

Throughout this section, we show several approaches used in the literature to pre-process time series by re-framing them into new representations for a further CNN implementation. Indeed, a raw time series needs to be converted into a set of fixed-length vector or matrix inputs before being fed into the CNN. Then, we discuss our data-level approach (of previous works [3, 4]) based on the Stockwell transform method.

3.1 Background on pre-processing methods for CNN

3.1.1 Basic pre-processing method: the sliding window

Given a sequence of values for a time series dataset, values at multiple time steps can be grouped to form an input vector while the output corresponds to a specific label given to this vector (generally provided by an expert). The use of time steps to predict the label (e.g., the class) is called the *Sliding Window* method and is explained in the algorithm below. The width of the sliding window L can be varied to include more or less previous time steps depending on the specificity of the dataset and the user preference.

Algorithm 1. Sliding window's algorithm

1. **procedure** *SlidingWindow*(T, L, s)
2. $i = 0; \ n = 0;$ // n is the number of frames/windows.
3. $F = [];$ // F is the set of extracted frames/windows.
4. **while** $i + L \leq length(T)$ **do** // L is the length of sliding window
5. $F[n] = T[i..(i + L - 1)];$ // T is the original time series data/sequence.
6. $i = i + s; \ n = n + 1;$ // s is the step.
7. **end while**
8. **return** F;
9. **end procedure**

3.1.2 Other pre-processing methods

Several research papers have focused mainly on applying some pre-processing to raw time series before being fed into the CNN. In this subsection, we present some important contributions which demonstrated that applying changes to the signals can further improve the CNN performance.

Several attempts have been made in order to encode raw time series as a matrix representation (e.g., 2D images) such as the Gramian Angular Field (GAF) [15], the Markov Transition Field (MTF) [15], Recurrence Plots (RP) [27], and stacked time series signals [29, 31], multivariate time series are treated as a 2D time-space input signals with one dimension denoting discrete time flows and the other corresponding to different channels of the multivariate time series.

Another type of data pre-processing based on applying transformation to data is performed in order to augment the data, thereby ensuring a better CNN training and thus a higher performance. For instance, the *window slicing* method [24] trains the CNN using slices of the time series input, then at test time classifies each slice of the test time series using CNN, and performs majority voting to output the predicted label. The *window warping* method [24] consists of warping a randomly selected slice of a time series by speeding it up or down, producing a transformed raw time series. Then, this latter is further converted into fixed-length input signals/instances via window slicing. Another attempt of augmenting time series is suggested in [3] where either small noise or smoothing is applied to the raw time series. Other transformations were also considered in [17] such as down-sampling to generate versions of a time series at different time scales, and spectral transformations in the frequency domain by adopting low frequency to remove noise from time series inputs.

Knowing that random noise and high-frequency perturbations present in the time series data can interfere tremendously with the learning process and that it is hard to capture useful features with the presence of noise in raw time series data, some works [5, 30] proposed to apply the Fast Fourier transform (FFT) and convert the raw time series into a set of frequency domain signals which serve as inputs for the CNN training process.

3.2 Stockwell transform

Instead of employing the FFT which is restricted to a predefined fixed window length, we choose to adopt the Stockwell transform (ST) as our preprocessing method for CNN training [3, 4]. In this section, the ST method is defined, its implementation on real world applications is detailed, and its experimental results are analyzed.

3.2.1 Methodology

The advantage of the ST over the FFT is its ability to adaptively capture spectral changes over time without windowing of data, resulting in a better time-frequency resolution for non-stationary signals [32]. To illustrate the ST method, let $h[l] = h(l{\bullet}T), l = 0, 1, \ldots, N-1$, be the samples of the continuous signal $h(t)$, where T is the sampling interval (i.e., the sampling interval of our sensor or measuring device). The discrete Fourier transform (DFT) can be written as,

$$H[m] = \sum_{l=0}^{N-1} h[l] e^{\frac{-i2\pi ml}{N}} \tag{1}$$

where m is the discrete frequency index $m = 0, 1, ..., N - 1$.

The discrete Stockwell transform (DST) is given by,

$$S[k, n] = \sum_{m=0}^{N-1} W[m]H[m+n]e^{\frac{i2\pi mk}{N}} \tag{2}$$

where k is the index for time translation and n is the index for frequency shift. The function $W[m] = e^{\frac{-2\pi^2 m^2}{n^2}}$ is the Gaussian window in the frequency domain.

Given a N-length signal, the DST coefficients are computed using the following steps:

1. Apply an N-point DFT to calculate the Fourier spectrum of the signal $H[m]$;

2. Multiply $H[m+n]$ with the Gaussian window function $W[m] = e^{\frac{-2\pi^2 m^2}{n^2}}$

3. For each fixed frequency shift $n = 0, 1, \cdots, \tau - 1$ (where τ is the number of frequency steps desired), apply an N-point inverse DFT to $W[m]H[m + n]$ in order to calculate the DST coefficients $S[k, n]$, where $k = 0, 1, ..., N - 1$;

Note that there is a DST coefficient calculated for every pair of $\langle k, n \rangle$ in the time-frequency domain. Therefore, the result of the DST is a complex matrix of size $\tau \times N$ where the rows represent the frequencies for every frequency shift index n and the columns are the time values of every time translation index k i.e., each column is the "local spectrum" for that point in time. This results in N instances, each instance being represented as a $\tau \times 1 \times 1$ matrix. If the time series is multivariate with D channels, D DST matrices will be generated, and each instance is represented as a $\tau \times 1 \times D$ matrix.

3.2.2 Experiments

3.2.2.1 Stereotypical motor movement (SMM) recognition task [4]

A SMM is defined as a repetitive movement which is regarded as one of the most apparent and relevant atypical behaviors present within children on the Autism Spectrum. Thus, detecting SMM behaviors can play a major role in the screening and therapy of ASD, thus potentially improving the lives of children in the spectrum.

Dataset. The SMM dataset used for training the CNN is derived from [33] and consists of raw time series of acceleration signals collected by three-axis wireless accelerometers (located at the torso, left wrist and right wrist) from six atypical (e.g., autistic) subjects in a longitudinal study. Activities including SMMs (body rocking, hand flapping, or simultaneous body rocking and hand flapping) and non-SMMs were engaged by subjects and were labeled (annotated) by an expert as SMM or non-SMM. Two to three sessions (9 to 39-min long) were recorded per participant, except subject 6 who was observed only once in Study 2.

Pre-processing. The data collection called "Study1" and "Study2" were recorded at a sampling frequency of 60 and 90 Hz respectively. So, to equalize the data, the 60 Hz signals are resampled and interpolated to 90 Hz. Next, data of both sensors go through a high pass filter with a cut-off frequency of 0.1 Hz in order to get rid of noise.

Afterwards, data is turned into fixed-length vector samples either in time-domain (using the sliding window) or in frequency-domain (using ST). Time-domain samples are obtained by segmenting raw data using a one second window (e.g., $L = 90$, see algorithm1) and 88.9% overlap between consecutive data segments (e.g. $s = 10$, see algorithm1), resulting in 90 time-point samples. And, knowing that three 3-axis acceleration signals are measured per accelerometer, an input sample will be a $90 \times 1 \times 9$ matrix with 9 denoting the number of channels ($9 = 3accelerometers \times 3coordinates$).

On the other hand, *frequency-domain* samples are obtained by deriving ST for every other 10th sample, and by selecting the proper best frequency range. Considering that 98% of the FFT amplitude of human activity frequencies is contained below 10 Hz, the ST frequencies are first chosen to be between 0 and 10 Hz, which yields bad CNN classification performance. And, after considering Goodwin's observation that almost all SMMs are contained within a frequency range of 1–3 Hz, we chose a new frequency range of 0–3 Hz which produced higher CNN classification performance. So, computing the ST generates multiple input samples (vectors of length 50), each containing the power of 50 frequencies ($\tau = 50$) in the range of 0–3 Hz. Thus, each extracted frequency-domain sample is a $50 \times 1 \times 9$ matrix.

CNN training. The purpose is to analyze intersession variability of different SMMs by training one CNN per domain (time or frequency domain) per subject per study. In other words, feature learning that is performed is specific to one domain (time and frequency), one subject i and one study j. The goal of this experiment is to build deep networks that are capable of recognizing SMMs across multiple sessions within the same atypical subject i. Training is conducted using k-fold cross-validation of an atypical subject i for a study j such that k is the number of sessions for which a participant was observed within each study, and every fold consists of data from a specific session. Time and frequency domain CNN architectures are composed of three and two sets of convolution, ReLU and pooling layers respectively with the number of filters set to {96, 192, 300} and {96,192} respectively, followed by a fully connected layer with 500 neurons. Training is performed for 10 to 40 epochs with the following hyper-parameters: a dropout of 0.5, a momentum of 0.9, a learning rate of 0.01, a weight decay of 0.0005, and a mini-batch size of 150.

3.2.2.2 Human activity recognition (HAR) task [3]

Dataset. The dataset used for HAR is the PUC dataset [34] which consists of 8 hours of human activities collected at a sampling frequency of 8 Hz by 4 tri-axial ADXL335 accelerometers located at the waist, left thigh, right ankle, and right arm. The activities are: sitting, standing, sitting down, standing up, and walking.

Pre-processing. The PUC data is further converted into time and frequency domain signals. In time-domain, a 1 s time window (e.g., $L = 8$) with 125 ms overlapping (e.g., $s = 1$) is employed to generate 8×1 time-domain samples. However, knowing that an 8×1 input matrix is not a vector long enough for training a CNN, signals are resampled from 8 to 50 using an antialiasing FIR low-pass filter and compensating for the delay introduced by the filter. The resultant time-domain input samples are $50 \times 1 \times 12$ matrix where 12 stands for the number of channels ($14accelerometers \times 3coordinates$). In frequency-domain, raw signals are resampled from 8 to 16 Hz; then, the ST is computed to obtain, for each input sample, the power of 50 frequencies in the range of 0–8 Hz, resulting in frequency-domain input samples of size $50 \times 1 \times 12$.

CNN training. In this experiment, one CNN is trained for each domain (time and frequency domain), with a 10-fold cross-validation. CNN architecture and parameters are set the same as in the SMM recognition task.

3.2.3 Results

Table 1 summarizes accuracy and F1-score results of CNN (in both time and frequency domains) for both the SMM recognition and HAR tasks. For SMM

	F1-scores SMM recognition												Accuracies HAR
	Study 1						Study 2						
	S1	S2	S3	S4	S5	S6	S1	S2	S3	S4	S5	Mean	
Time-domain CNN	91.23	76.76	84.95	93.38	86.41	95.11	95.97	75.67	60.17	91.68	82.55	84.90	**99.90**
Frequency-domain CNN	**96.54**	**78.41**	**93.62**	**96.46**	**95.74**	**98.58**	**96.07**	**95.27**	**85.03**	**98.03**	**93.88**	**93.42**	95.98

Table 1.
Performance rates of time-, and frequency-domain CNNs for the SMM recognition (in terms of F1-score) and Human Activity Recognition referred to as HAR (in terms of accuracy). Highest rates are in bold.

recognition, as opposed to other subjects, Subject 6 within Study 2 had only one session recorded; thus, no CNN was trained for this subject. We observe that all frequency-domain CNNs (of all subjects in all studies) perform better than time-domain CNNs by 8.52% (in terms of the mean F1-score). This suggests that ST eliminates all noisy information and thus helps the CNN capture meaningful features.

However, as opposed to these results, comparing results of time and frequency domain CNNs on the Human Activity Recognition (HAR) task demonstrates the efficiency of time over frequency by 3.92% in terms of accuracy (as shown in **Table 1**). These contradictory results can be explained by the difference in the chosen ST frequency range for SMM recognition and that of HAR. Indeed, in SMM recognition, the frequency range of the ST was carefully chosen to cover almost all SMMs (0–3 Hz), resulting in optimal frequency-domain samples (containing full and noise-free information) which produced better CNN parameters. Meanwhile, the ST frequency range for HAR (0–8 Hz) may be a short/small range which generated frequency-domain samples that may have lost relevant information. Indeed human activity frequencies fall between 0 and 20 Hz (with 98% of the FFT amplitude contained below 10 Hz). Thus, in order to train CNNs with frequency-domain signals, it is necessary to analyze raw time series to come up with the proper ST frequency range which covers all valuable information needed for the recognition task.

4. Algorithm-level approach

4.1 Background on convolutional neural networks

4.1.1 Definition

CNNs were developed with the idea of local connectivity. Each node is connected only to a local region in the input. The local connectivity is achieved by replacing the weighted sums from the neural network with convolutions. In each layer of the CNN, the input is convolved with the weight matrix (e.g., the filter) to create a feature map. As opposed to regular neural networks, all the values in the output feature map share the same weights so that all nodes in the output detect exactly the same pattern. The local connectivity and shared weights aspect of CNNs reduce the total number of learnable parameters, resulting in more efficient training and learning in each layer a weight matrix which is capable of capturing the necessary, translation-invariant features from the input.

4.1.2 CNN structure

The input to a convolutional layer is usually taken to be three-dimensional: the height, weight and number of channels. In the first layer this input is convolved with a set of M_1 three-dimensional filters applied over all the input channels. In our case, we consider a one-dimensional time series $x = (x_t)_{t=0}^{N-1}$. Given a classification task and a model with parameter values w, the task for a classifier is to output the predicted class $\hat{y}$ based on the input time series, $x(0), \ldots, x(t)$.

The output feature map from the first convolutional layer is then given by convolving each filter w_h^1 for $h = 1, \ldots, M_1$ with the input:

$$a^1(i,h) = \left(w_h^1 * x\right)(i) = \sum_{j=-\infty}^{\infty} w_h^1(j)x(i-j) \tag{3}$$

where $w_h^1 \in R^{1\times k\times 1}$ and $a^1 \in R^{1\times N-k+1\times M_1}$, i is the index of the feature map at the second dimension ($i = 1, \ldots, N-k+1$) and h is the index of the feature map at the third dimension ($h = 1, \ldots, M_1$). Note that since the number of input channels in this case is one, the weight matrix also has only one channel. Similar to the feedforward neural network, this output is then passed through the non-linearity $h(\cdot)$ to give $f^1 = h(a^1)$.

In each subsequent layer $l = 2, \ldots, L$, the input feature map $f^{l-1} \in R^{1\times N_{l-1}\times M_{l-1}}$, where $1 \times N_{l-1} \times M_{l-1}$ is the size of the output filter map from the previous convolution with $N_{l-1} = N_{l-2} - k + 1$, is convolved with a set of M_l filters $w_h^l \in R^{1\times k\times M_{l-1}}$, $h = 1, \ldots, M_l$, to create a feature map $a^l \in R^{1\times N_l\times M_l}$,

$$a^l(i,h) = \left(w_h^l * f^{l-1}\right)(i) = \sum_{j=-\infty}^{\infty} \sum_{m=1}^{M_{l-1}} w_h^1(j,m) f^{l-1}(i-j,m) \tag{4}$$

This output is then passed through the non-linearity to give $f^l = h(a^l)$. The filter size parameter k thus controls the receptive field of each output node. Without zero padding, in every layer the convolution output has width $N_l = N_{l-1} - k + 1$ for $l = 1, .., L$. Since all the elements in the feature map share the same weights this allows for features to be detected in a time-invariant manner, while at the same time it reduces the number of trainable parameters.

The output is then fed into a pooling layer (usually a max-pooling layer) which acts as a subsampling layer. The output map $p^l(h)$ of a feature map h is achieved by computing maximum values over nearby inputs of the feature map as follows:

$$p^l(i,h) = \max_{r\in R}\left(f^l(i \times T + r, h)\right) \tag{5}$$

where R is the pooling size, T is the pooling stride, and i is the index of the resultant feature map at the second dimension.

Multiple convolution, ReLU and pooling layers can be stacked on top of one another to form a deep CNN architecture. Then, the output of these layers is fed into a fully connected layer and an activation layer. The output of the network after L layers will thus be the matrix f^L. Depending on what we want our model to learn, the weights in the model are trained to minimize the error between the output from the network f^L and the true output we are interested in, which is often denoted as the objective function (loss function). For instance, a softmax layer can be applied on top, followed by the entropy cost function which is an objective function computed based on the true labels of training instances and probabilistic outputs of softmax function.

4.2 CNN with the adaptive convolutional filter approach

In previous CNN works, several attempts have been made to extract the most relevant/meaningful features using different CNN architectures. While works [17, 24] transformed the time series signals (by applying down-sampling, slicing, or warping) so as to help the convolutional filters (especially the 1st convolutional layer filters) capture entire peaks (i.e., whole peaks) and fluctuations within the signals, the work of [18] proposed to keep time series data unchanged and rather feed them into three branches, each having a different 1st convolutional filter size, in order to capture the whole fluctuations within signals. An alternative is to find an adaptive 1st convolutional layer filter which has the most optimal size and is able to capture most of entire peaks present in the input signals. By obtaining the most appropriate 1st convolutional layer filter, there will be no need to apply multiple branches with different 1st convolutional layer filter sizes, and no need to apply transformations such as down-sampling, slicing and warping, thus requiring less computational resources. The question of how to compute this adaptive 1st convolutional layer filter is addressed in [4]. In this section, we will discuss the approach based on the adaptive 1st convolutional layer filter. Next, to prove the efficiency of this/our approach, an application on SMM recognition is conducted and results are analyzed.

4.2.1 Methodology

In CNNs, multiple hyper-parameters are to be chosen carefully in order to retrieve the best classification rate, including model hyper-parameters which define the CNN architecture, and optimization hyper-parameters such as the loss function, learning rate, etc. Model Hyper-parameter values are generally chosen based on the literature and on the trial-and-error process (through running experiments with multiple values). A conventional approach is to start with a CNN architecture which has already been adopted in a similar domain to ours, and then update hyper-parameters by experimentation.

In our study, we focus on the convolutional layer filter (also known as "Receptive field"). Conventionally, the 1st convolutional layer filter has one of the following sizes: 3×3, 5×5, 7×7 and 10×10, where small filter sizes capture very fine details of the input while large ones leave out minute details in the input. After all, the goal is to set the receptive field size such that the filters detect complete features (e.g., entire variations) within an object (which could be edges, colors or textures within an image, or peaks within a signal). Choosing a small 1st layer filter is good for capturing a short variation (e.g., a signal peak) within a time series input signal, but may capture only a slice of this variation by convolving only part of it, thus failing to detect the whole fluctuation within the signal. Conversely, a relatively large filter may convolve multiple signal peaks at once and therefore no fluctuation is detected either. So, the choice of the proper 1st layer receptive field size is crucial for find good CNN features and for maximizing recognition performance. In that sense, we need 1st layer filter that suits the input signals and variations present within them. In other words, we need to find the best filter size which convolves most of the entire signal peak within the input signals. To this end, it is necessary to find the optimal length of all signal peaks present within input signals. To do so, we apply *sampling*, a statistical procedure that uses characteristics of the sample (i.e., sample peak lengths taken from randomly selected signals) to estimate the characteristics of the population (i.e., the optimal peak length of all signals). The sample statistic can be the mean, median or mode.

Given a population of signals with mean μ and n random values x (n being sufficiently large: $n \geq 30$) sampled from the population with replacement, the mean $E(x)$ is a point estimate of μ and $E(x) = \mu$ where $E(x) = \frac{1}{n}\sum_{i=1}^{n} x_i$. However, using the sample mean $E(x)$ as the optimal length of peaks within time series signals (in time- and frequency-domains) and as the size of the 1st convolutional layer filter gives poor CNN performance since it is influenced by outliers or extremes values (i.e., by signals peaks whose length are either too small and too big). In this case, we use the sample median.

For the sample median $M_e(x)$ to be a point estimate of the population median M_e (with a small bias), the distribution of the sample values x should be normally distributed. Nonetheless, the asymptotic distribution of the sample median in the classical definition is well-known to be normal for absolutely continuous distributions only and not for discrete distributions (such as time series). A solution to this problem is to employ the definition of the sample median based on mid-distribution functions [35] which proves that the sample median has an asymptotically normal distribution, meaning that $M_e(x) \approx M_e$. Then, computing the sample median of signal peak lengths will give us the optimal size of the 1st convolutional layer filter.

4.2.2 Experiments

Experiments are conducted on the SMM Recognition task. The dataset and experimental setup are the same as in Section 3.2.2. The inputs used are either time-domains acceleration signals of size $90 \times 1 \times 9$ for time-domain CNN training or frequency-domain signals of size $50 \times 1 \times 9$ for frequency-domain CNN training. The goal is to find the optimal size of the 1st convolutional layer filter for both time-domain and frequency-domain CNNs.

As explained in the methodology, the first step to determine the size of the 1st convolutional layer filter is to collect 30 random signals (for each of the time and frequency domain SMM signals) that contain at least one peak and to randomly pick 30 peaks from these signals. Histograms (a) and (b) of **Figure 1** represent frequency distributions of the 30 peak lengths for time and frequency domain signals respectively. Afterwards, the computed time and frequency domain medians (9 and 10 respectively) are applied as the optimal size of the 1st convolutional layer filter for the time and frequency domain CNNs respectively.

4.2.3 Results

In order to prove the efficiency of this adaptive 1st convolutional layer filter approach, we run experiments on different time and frequency domain CNN architectures by varying the size of the 1st convolutional layer filter between 7 and 11 across both architectures. Performance rates in terms of the F1-score metric are displayed in **Figure 2**. In time-domain, an increase in the size of the 1st convolutional layer filter from 7 ($\sim$ a time span of 0.078 s) to 9 ($\sim$ a time span of 0.1 s) results in an increase of 3.26%, while an increase of the filter size from 9 to 10 ($\sim$ a time span of 0.11 s) and 11 ($\sim$ a time span of 0.12 s) diminishes the performance of the network. Therefore, the most optimal size of the 1st convolutional filter is equal to the sample median of signal peak lengths, suggesting that 0.1 is the best time span of the 1st convolutional layer to retrieve the whole acceleration peaks and the best acceleration changes. Similarly, in frequency domain, the 1st convolutional layer kernel yielding the highest F1-score is the one with size 10, which is simply the sample median ($M_e(x) = 10$). Thus, these results confirm the

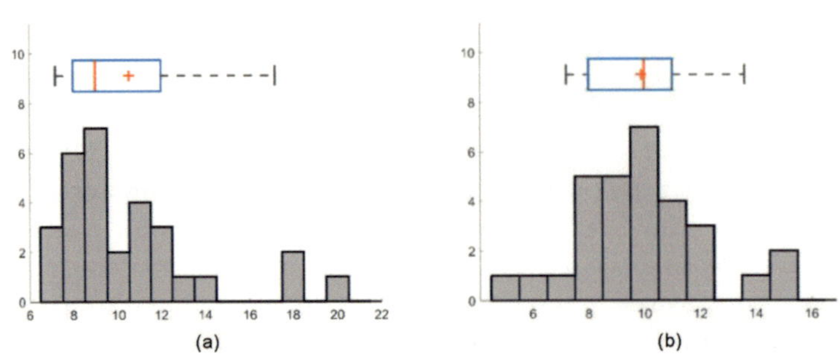

Figure 1.
(a) and (b) Histograms and box plots of the frequency distribution of 30 peak lengths present within 30 randomly selected time and frequency domain signals respectively.

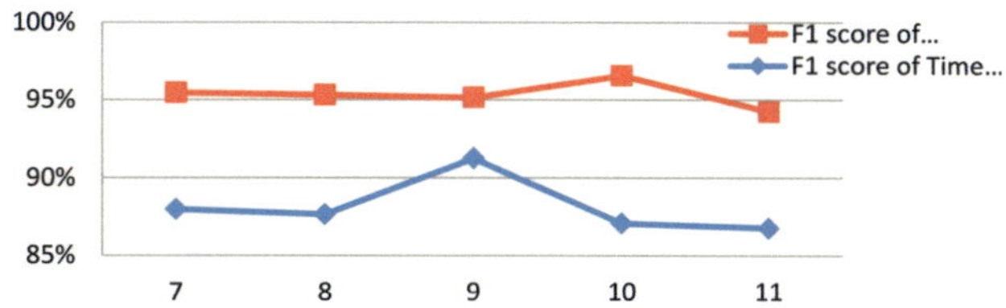

Figure 2.
Effect of the size of 1st convolutional layer kernel on SMM recognition performance.

	Study 1						Study 2					Mean
	S1	S2	S3	S4	S5	S6	S1	S2	S3	S4	S5	
CNN-Rad [30]	71	73	70	92	68	94	68	22	2	77	75	64.73
Time-domain CNN	91.23	76.76	84.95	93.38	86.41	95.11	95.97	75.67	60.17	91.68	82.55	84.90

Table 2.
Comparative results (F1-scores) between the CNN using the adaptive 1st convolutional filter approach and the CNN of Rad et al. [30].

superiority of the adaptive 1st convolutional layer filter approach for both time and frequency domain signals.

Furthermore, another way to show the efficiency of this adaptive 1st convolutional layer filter approach is to compare the performance of our time-domain CNN with the CNN of Rad et al. [30] which was trained on the same dataset as ours (in time-domain). **Table 2** displays F1-score results of CNNs trained per subject and per study using the optimal 1st convolutional layer filter size (denoted as "Time-domain CNN") and using the architecture of [30] (referred to as CNN-Rad). These results suggest that our time-domain CNN performs 20.17% higher in overall than the CNN of [30] and confirms the efficiency of the adaptive convolutional layer.

4.3 CNN approach for tasks with limited annotated data

CNNs have so far yielded outstanding performance in several time series applications. However, this deep learning technique is a data driven approach i.e., a supervised machine learning algorithm that requires excessive amount of labeled (e.g., annotated) data for proper training and for a good convergence of parameters. Although in recent years several labeled datasets have become available, some fields such as medicine experience a lack of annotated data as manually annotating a large set requires human expertise and is time consuming. For instance, labeling acceleration signals of autistic children as SMMs or non-SMMs requires knowledge of a specialist. The conventional approach to deal with this kind of problem is to perform data augmentation by applying transformations to the existing data, as shown in Section 3.1.2. Data augmentation achieves slightly better time series classification rates but still the CNN is prone to overfitting. In this section, we present another solution to this problem, a "knowledge transfer" framework which is a global, fast and light-weight framework that combines the transfer learning technique with an SVM classifier. Afterwards, this technique is further implemented on another type of SMM recognition task, which consists of recognizing SMMs across different atypical subjects rather than recognizing SMMs across multiple sessions within one subject (as performed in experiments of Sections 3.2.2 and 4.2.2).

4.3.1 Methodology

Transfer learning is a machine learning technique where a model trained on one task (a source domain) is re-purposed on a second related task (a target domain). Transfer learning is popular in deep learning, including Convolutional Neural Networks, given the enormous resources required to train deep learning models or the large and challenging datasets on which deep learning models are trained. For a CNN, given a source domain with a source learning task and a target domain with a target learning task (task of interest), transfer learning aims to improve learning of the target predictive function using the knowledge in the source domain which is the pre-trained CNN model containing features (parameters or weights) learned from the source domain task. This process works if these source domain features are general, meaningful and suitable to the target task. The pre-trained model can then be used as the starting point for a model on the target task. This may involve using all or parts of the pre-trained CNN model, depending on the modeling technique used. Accordingly, the questions that arise are: (i) which source learning task should be used for pre-training the CNN model given a target learning task, and (ii) which parts (e.g., learned features) of this model are common between the source and target learning tasks.

An answer to the first question is to propose two source learning tasks. One source learning task is chosen to be very close and similar to the target learning task. And, if this source learning task lacks annotated data, then another source learning task is introduced which is chosen to be different but related to the target learning task.

A solution to the second problematic is to assume that features shared across the source and target tasks correspond to low- and mid-level information (e.g., fine details or local patterns) contained within inputs of both tasks, whereas the unshared features are the high-level information (e.g., global patterns) contained within inputs. And, knowing that training a CNNs produces learned low-, mid- and high-level features located at (contained within) the first, intermediate and last hidden layers respectively, we therefore assume that the features shared between the source and target tasks are contained within the first and intermediate CNN layers while features distinguishing one task from the other are contained within the last CNN layer. For instance, in image classification, as the CNN learns low-level features (Gabor filters such as edges, corners) through the first hidden layers, mid-level features (squares, circles, etc.) through intermediates hidden layers, and high-level features (faces, text, etc.) through last hidden layers, scene recognition (source learning task) and object recognition (target learning task) will have the same first and intermediate layers' weights but different last layer weights. In time series, considering human activities where every activity is a combination of several basic continuous movements, with basic continuous movements corresponding to the smooth signals and the transitions or combinations among these movements causing the significant/salient changes of signal values, the purpose of the CNN will be to capture basic continuous movements through its low- and mid-level parameters (first and intermediate hidden layers) and the salience of the combination of basic movements through its high-level parameters (last hidden layers). Therefore, as an example, the CNN trained on recognizing basic human activities such as sitting, standing and walking (source learning task), and the one trained on recognizing SMMs (target learning task) will both have the same first and intermediate layer weights and different high layer weights. Another example is the CNN trained on SMMs of an atypical subject (source task) and the one trained on SMMs of another atypical subject (target task) which will have common first and intermediate hidden layers' weights and different last hidden layer weights, due to the inter-subject variability across atypical subjects.

In that sense, we propose a "Transfer learning with SVM read-out" framework which is composed of two parts: (i) the first part having first and intermediate layers' weights of a CNN already pre-trained on a source learning task, (the last CNN layer being discarded), and (ii) the second part composed of a support vector machine (SVM) classifier with RBF kernel which is connected to the end of the first part. Then, we feed the entire training dataset of the target task into this framework in order to train the SVM parameters. As opposed to training a CNN on the target task which requires updating all hidden layers' weights for several iterations using a large training set for all these weights to converge, our framework computes weights of the last layer(s) only, in one iteration only. Moreover the advantage of using SVM as the classifier is that it is fast and generally performs well on small training set since it only relies on the support vectors, which are the training samples that lay exactly on the hyperplanes used to define the margin. In addition, SVMs have the powerful RBF kernel, which allows to map the data to a very high dimension space in which the data can be separable by a hyperplane, hence guaranteeing convergence. Hence, our framework can be regarded as a global, fast and light-weight technique for time series classification where the target task has limited annotated/labeled data.

4.3.2 Experiments

We conduct this experiment again on the SMM recognition task. However, we will perform SMM recognition across multiple atypical subjects as opposed to SMM recognition within subjects which was developed in experiments of Sections 3.2. and 4.2. Indeed, due to the inter-subject variability of SMMs, a CNN trained on movements of an atypical subject i performs badly on detecting SMMs of another atypical subject and therefore cannot be applied on SMMs other than subject i's SMMs. Indeed, testing one of the trained CNNs of experiment 4.2.2 (let us say the trained CNN of subject i study j) on SMMs of a subject other than subject i produces a very low F1-score with less than 30%. This implies that CNN features learned from SMMs of one subject differ from the ones learned from another subject and that they are not general enough to detect SMMs of another subject. So, instead of training a CNN for each atypical subject individually (Sections 3.2.2 and 4.2.2), we use the "transfer learning with SVM read-out" framework for the detection of SMMs across subjects. Through this experiment, our goal is to prove that this framework is a global, fast and light-weight technique for time series classification tasks which experience a lack of labeled data.

The target learning task will be the recognition of SMMs of subject i, while the source learning task will be either a task close to the target task such as the recognition of SMMs of multiple subjects other than i, or a task different but related to the target task such as the recognition of basic human activities. Running the "TL SVM" framework with the former and the latter target tasks will be denoted as "*TL SVM similar domains*" and "*TL SVM across domains*" respectively. Through this experiment, we will also show that these chosen source learning tasks contribute in generating CNN features that are general/global enough to recognize SMMs of any new atypical subject.

Datasets. The dataset used for the target learning task is the same SMM dataset used in Section 3.2.2. The dataset used for the source domain of the "*TL SVM similar domains*" experiment is also the SMM dataset, whereas the one used for the source domain of the "TL SVM across domains" experiment is the PUC dataset which is described in Section 3.2.2.

When using the SMM dataset in the target and source learning tasks, we do not take into consideration signals of all accelerometers/sensors (torso, right and left wrist) but rather signals of the torso sensor, resulting in input samples with 3 channels instead of 9. So, with torso measurements only, the only stereotypical movements that could be captured are the rock and flap-rock SMMs (and no flap SMMs). Accordingly, only rock and flap-rock SMM instances will be used as inputs in this experiment.

When using the PUC dataset for the source learning task, only the waist accelerometer (waist being next to torso) is taken into account since the other accelerometers (located at the thigh, ankle and arm) will not be relevant to the SMM recognition task during transfer learning. We consider the waist location to be equivalent to the torso location so that the CNN pre-trained on the source learning dataset can further be transferred to the target learning task (SMM recognition). Accordingly, input instances will have 3 channels instead of 12.

Pre-processing. The pre-processing phase is the same as in Section 3.2.2.

Experimental setup. In experiments below, the architecture of the CNN model in time domain and frequency domain as well as training parameters are similar to the ones in Section 3.2.2. In addition, the target learning task consists of SMM recognition of a target subject i of study j where $i \in [1, 6]$ and $j \in [1, 2]$. Accordingly, one "transfer learning with SVM" framework will be run per domain (time or frequency domain) per subject per study. The training and testing sets of subject i (study j) are selected using the same k-fold cross-validation used in Section 3.2.2. However only a subset of the training set (10,000–30,000 instances) is used, where 2000 SMM instances are randomly selected from the overall training set for training.

In order to perform SMM recognition on a target subject using transfer knowledge from SMMs of other subjects, the following steps are performed in time and frequency domains for each study i within each study j:

- **Step 1:** we train a randomly initialized CNN in both time and frequency domains, for 5–15 epochs, using: (i) SMM instances of all 6 atypical subjects within study j except subject i for "*TL SVM similar domains*" framework, and (ii) basic human activities' instances for "TL SVM across domains" framework. This process results in a pre-trained CNN model.

- **Step 2:** we reuse all layers of this CNN except the last layer (which is a fully connected layer) which is removed and replaced by the SVM classifier. The SVM of the transfer learning framework is trained using a small subset of subject i's training data (i.e., 2000 SMM samples), which results in learned high-level features. Then, the remaining SMMs of subject i are implemented for testing the framework. Knowing that the input consists of only a subset of the original training dataset of subject i, we choose to run the SVM for 5 runs, with 2000 randomly selected samples in each run. In such a way, by aggregating F1-scores of the 5 runs, we provide more realistic results. This procedure is applied on every domain (time and frequency) on every subject i in every study j.

4.3.3 Results

"TL SVM similar domains". As depicted in **Table 3**, this framework (combining part of the pre-trained CNN with an SVM) is able to identify SMMs at a mean F1-score of 74.50 and 91.81% for time and frequency domains respectively. As opposed to the technique of directly applying the pre-trained CNN for classification which fails to recognize SMMs, "*TL SVM similar domains*" framework is able to capture relevant features for the recognition of SMMs across subjects. Thus, we can infer that low and mid-level SMM features share the same information from one subject to another and that "*TL SVM similar domains*" can be used as a global framework to identify SMMs of any new atypical subject. Furthermore, low- and mid-level features captured from a source learning task can be employed as low- and mid-level features of a target learning task close to the source task.

"TL SVM across domains". Training this framework produces satisfying results with a mean score of 72.29 and 79.78% in time and frequency domains respectively (**Table 3**). So, fixing low and mid-level features to features of basic movements and adjusting only the high-level features by an SVM seems to give satisfying classification results, which confirms that our framework has engaged feature detectors for finding stereotypical movements in signals. These results, especially the frequency-domains results, indicate that: (i) connecting low- and mid-level features of basic movements to an SVM classifier then feeding in 2000 instances for training the SVM generates a global framework which holds relevant and general representation that adapts to SMMs of any new atypical subject i, and (ii) both human and stereotypical movements may share low and mid-level features in common, suggesting that low- and mid-level information learned from a source target task by a CNN model can be directly applied as low- and mid-level features for a target learning task different but related to the source learning task, especially when there is a lack of labeled data within the target learning task.

Moreover, both our techniques are compared against the following methods:

i. The "*CNN with few data*" technique which consists of training a CNN in time and frequency domains with randomly initialized weights using the same target training data as the ones of "*TL SVM similar domains*" framework (i.e., 2000 SMM instances of the target subject i). The difference between this CNN and the CNN of Section 3.2.2 is that less data is used for training (2000 versus 10,000–30,000 training instances), only torso sensor measurements are applied in the former (compared to torso, right and left wrist sensor measurements in the latter), and only rock and flap-rock SMM instances are considered in the former (compared to rocking, flap-rock and flap SMM instances in the latter). We refer to this technique as "*CNN few data*".

ii. The "*transfer learning with full fine-tuning*" technique (referred to as "*TL full fine-tuning*") consists of identifying SMMs of subject i within study j by first training a CNN in time and frequency domains for 5–15 epochs using SMM instances of all 6 atypical subjects within study j except subject i (as in Step 1 of training "*TL SVM similar domains*" framework), then by fine-tuning (e.g., updating) weights of all CNN layers using same target training data.

iii. The "*transfer learning with limited fine-tuning*" technique (denoted as "*TL limited fine-tuning*") is the same as "*transfer learning with full fine-tuning*" except that the fine-tuning process is effective only on weights of the last CNN layer L while weights of other layers $1, \ldots, L-1$ are unchanged.

Results and properties of the three techniques are depicted in **Tables 3** and **4** respectively. From these results and properties, the following observations can be made:

- "*TL SVM similar domains*" framework performs higher than the three frameworks "*CNN few data*", "*TL full fine-tuning*" and "*TL limited fine-tuning*" in both time- and frequency-domain. This can be explained by the nature of the training process of the three frameworks, which relies on updating

Approaches	Study 1						Study 2					Mean
	S1	S2	S3	S4	S5	S6	S1	S2	S3	S4	S5	
Time domain												
CNN few data	72.02	62.31	52.98	88.47	60.61	88.86	80.90	53.28	16.00	82.66	75.82	66.72
TL full fine-tuning	**75.73**	71.31	**59.04**	91.67	59.47	91.89	85.66	**63.84**	38.57	92.24	**82.31**	73.79
TL limited fine-tuning	75.44	56.50	50.86	**91.74**	**63.86**	**93.11**	**85.88**	62.62	27.14	**93.63**	81.16	71.09
TL SVM similar domains	75.37	**76.44**	56.53	**91.74**	63.37	92.76	84.86	62.97	**41.60**	93.32	80.55	**74.50**
TL SVM across domains	71.66	74.40	66.80	90.69	61.87	92.19	81.35	58.13	35.66	88.44	73.98	72.29
Frequency domain												
CNN few data	76.64	96.55	63.44	93.13	82.58	94.61	84.94	76.42	29.51	93.66	83.41	79.54
TL full fine-tuning	88.51	**97.22**	**88.15**	97.53	**91.29**	98.26	92.17	88.19	52.17	96.59	91.98	89.28
TL limited fine-tuning	87.98	94.59	62.70	**97.57**	87.94	98.36	91.62	87.08	40.00	97.82	90.82	85.14
TL SVM similar domain	**90.54**	**97.22**	83.86	95.24	86.19	**98.45**	**92.71**	**90.49**	**84.99**	**97.99**	**92.22**	**91.81**
TL SVM across domains	74.50	91.56	43.77	93.11	76.03	94.2	85.16	74.67	66.98	93.66	83.99	79.78

Table 3.
Results of CNN approaches used in this experiment per domain (time or frequency) per subject, per study. Highest rates are in bold.

Approaches	# parameters updated for one pass (1 batch)	# batches per iteration	# iterations (epochs)	Implementation on Android device
CNN few data	1.2e + 06 (time-domain) 7.1e + 05 (frequency-domain)	14 (2000/150)	20–35 (time-domain) 55–85 (frequency-domain)	No (too much memory consumption)
TL full fine-tuning	1.2e + 06 (time-domain) 7.1e + 05 (frequency-domain)	14 (2000/150)	5–15	No (too much memory consumption)
TL limited fine-tuning	1000 (500*2)	14 (2000/150)	5–15	No (hard to run back-propagation on mobile devices)
TL SVM (similar domains and across domains)	500	1	1	Yes (easy to train SVM on mobile devices)

Table 4.
Properties and resources used for the different techniques implemented in the experiment.

parameters using backpropagation. And, knowing that backpropagation requires abundant data for proper training, a lack of training data (2000 SMM instances) pushes the three frameworks to overfit and be less efficient.

- "*TL SVM similar domains*" and "*TL SVM across domains*" architectures perform better than "*CNN few data*" by 7.78 and 5.57% respectively in time-domain and by 12.27 and 0.24% respectively in frequency-domain. Therefore, both architectures engage in capturing more general features than "*CNN few data*". In terms of resources, one advantage of the two architectures over "*CNN few data*" is that the former converge much faster than the latter. Indeed, the former require 5–15 epochs (in both time and frequency domains) for full convergence while the latter needs 20–35 epochs and 55–85 epochs in time- and frequency-domain respectively for full convergence, as shown in **Table 4**. Another advantage resides in the number of parameters that have to be learned, which is 500 for the former (in both time and frequency domains) versus 1.2e + 06 and 7.1e + 05 for the latter in time and frequency domain respectively (**Table 4**). Hence, as opposed to "*TL full fine-tuning*" and "*TL limited fine-tuning*" frameworks, the "TL SVM" can be regarded as a global, fast and light-weight framework for SMM recognition across subjects.

- "*TL full fine-tuning*" has a slightly higher performance than "*TL limited fine-tuning*" by 2.71 and 4.14% in time- and frequency-domain respectively, suggesting that fine-tuning weights of layers 1, …, $L - 1$ (where L is the number of CNN layers) is unnecessary since it does not improve SMM recognition significantly. This confirms the earliest assumption that atypical subjects have similar low- and mid-level features but different high-level features.

- "*TL SVM across domains*" framework yields a lower performance than "*TL SVM similar domains*" by 2.21% and 12.02% in time- and frequency-domain respectively. This implies the superiority in the SMM recognition task of low and mid-level features learned from SMMs over the ones learned from basic human movements. However, the latter features are more global. In time-domain, the small rate difference (2.21%) between "*TL SVM across domains*" and "*TL SVM similar domains*" suggests that the low- and mid-level feature space generated by human activities shares common details with the one generated by movements of specific atypical subjects. This is not the case for frequency-domain series, which can be explained by the difference in the frequency range between human activities and SMMs. Indeed, the FFT amplitude of human activities is contained below 10 Hz, pre-training the CNN on human activity frequency-signals from 0 to 3 Hz and not from 0 to 10 Hz results in imperfect human activity features which, combined with the SVM, do not seem to yield good classification results on the recognition of SMMs. If we were to have a new target learning task whose data signals are within the same frequency range as data signals of the source learning task, then "*TL SVM across domains*" would have achieved the same performance as "*TL SVM similar domains*".

- One advantage of "*TL SVM similar domains*" and "*TL SVM across domains*" is that they can be implemented in Android portable devices, as shown in **Table 4**. Indeed, an expert could receive continuous acceleration signals from the torso accelerometer of a subject, and label them on the fly (as SMM/non-SMM) as the subject performs his activities/movements. This results in annotated time series which are then preprocessed and fed into either "*TL SVM similar domains*" or "*TL SVM across domains*" for training. A one-minute recording of these signals is sufficient to train one of the two frameworks. Afterwards, this framework is ready to use for recognizing further SMMs on that same subject.

5. Conclusion

Time series pose important challenges to existing approaches which perform predictive modeling for classification tasks. In this paper we present a review on our previous works. Our contributions are aggregated into two categories: data-level and algorithm-level approaches. Our data-level approach consists of encoding time series using STin order to produce noise-free input signals which offer a more efficient CNN training. At the algorithm level, one approach is the adaptive convolutional layer filter approach which consists of determining the size of the filter based on an analysis of the input time series signals and fluctuations present within them. Indeed, choosing the proper 1st layer filter generates features maps which are more informative about the input signals and which capture the whole peaks within input signals. Furthermore, "*TL SVM similar domains*" and "*TL SVM across domains*" are algorithm-level approaches dealing with tasks with limited annotated data, which are regarded as two global, fast and light-weight techniques for these kinds of tasks. These two CNN approaches generate features general and global enough to recognize time series of the target learning task, given time series of a source learning task that is similar or different but related to the target learning task. All these approaches were implemented on the recognition of human activities, including normal activities performed by typical subjects and disorder-based activities performed by atypical subjects (such as SMMs of autistic subjects). Experimental results have showed the superiority of our techniques and their ability to extract relevant features from time series inputs. As a perspective, knowing that time series datasets often contain outliers either due to noisy time series or mislabeled time series (e.g. incorrect labels), we aim at studying a robust CNN that is insensitive to outliers. As opposed to our data-level CNN technique (mentioned in this paper) whose goal is to eliminate noise from time series, this robust CNN is an algorithm-level technique with acts at the level of loss functions by controlling high error values caused by outliers.

Conflict of interest

The authors declare that they have no conflicts of interest.

Author details

Lamyaa Sadouk
Faculty of Science and Technology, University Hassan 1st, Settat, Morocco

*Address all correspondence to: lamyaa.sadouk@gmail.com

References

[1] Kaya H, Gunduz-oguducu S. A distance based time series classification framework. Information Systems. 2015; **51**(C):27-42

[2] Wang S, Liu P, She MFH, et al. Bag-of-words representation for biomedical time series classification. Biomedical Signal Processing and Control. 2013; **8**(6):634-644

[3] Sadouk L, Gadi T, Essoufi EH. Convolutional neural networks for human activity recognition in time and frequency-domain. In: Proceedings of ICRTS. 2017. pp. 485-496

[4] Sadouk L, Gadi T, Essoufi EH. A novel deep learning approach for recognizing stereotypical motor movements within and across subjects on the autism spectrum disorder. Computational Intelligence and Neuroscience. 2018:16. DOI: 10.1155/2018/7186762. Article ID 7186762

[5] Abdel-Hamid O, Deng L, Yu D. Exploring convolutional neural network structures and optimization techniques for speech recognition. Interspeech. 2013;**2013**:1173-1175

[6] Abdel-Hamid O, Mohamed AR, Jiang H, Penn G. Applying convolutional neural networks concepts to hybrid nn-hmm model for speech recognition. In: Proceedings of ICASSP IEEE. 2012. pp. 4277-4280

[7] Kini BV, Sekhar CC. Large margin mixture of AR models for time series classification. Applied Soft Computing. 2013;**13**(1):361-371

[8] Antonucci A, De Rosa R, Giusti A, et al. Robust classification of multivariate time series by imprecise hidden Markov models. International Journal of Approximate Reasoning. 2015;**56**(B):249-263

[9] Rakthanmanon T, Campana B, Mueen A, et al. Addressing big data time series: Mining trillions of time series subsequences under dynamic time warping. ACM Transactions on Knowledge Discovery from Data. 2013; **7**(3):1-31

[10] Jeong Y, Jeong MK, Omitaomu OA. Weighted dynamic time warping for time series classification. Pattern Recognition. 2011;**44**(9):2231-2240

[11] Schäfer P. The BOSS is concerned with time series classification in the presence of noise. Data Mining and Knowledge Discovery. 2014;**29**(6):1505-1530

[12] Bailly A et al. Dense bag-of-temporal-SIFT-words for time series classification. In: Lecture Notes in Artificial Intelligence. 2016

[13] Chen Z, Zuo W, Hu Q, Lin L. Kernel sparse representation for time series classification. Information Sciences. 2015;**292**:15-26

[14] Hills J, Lines J, Baranauskas E, et al. Classification of time series by shapelet transformation. Data Mining and Knowledge Discovery. 2014;**28**(4):851-881

[15] Wang Z, Oates T. Encoding time series as images for visual inspection and classification using tiled convolutional neural networks. In: Workshops AAAI. 2015

[16] Ngiam J, Chen Z, Chia D, Koh PW, Le QV, Ng AY. Tiled convolutional neural networks. In: Advances in Neural Information Processing Systems. 2010:1279-1287

[17] Cui Z, Chen W, Chen Y. Multi-scale convolutional neural networks for time series classification. 2016. arXiv preprint arXiv:1603.06995

[18] Wenlin W et al. Earliness-aware deep convolutional networks for early time series classification. 2016. arXiv preprint arXiv:1611.04578

[19] Wang Z, Yan W, Oates T. Time series classification from scratch with deep neural networks: A strong baseline. In: Proceedings of the IEEE IJCNN. 2017. pp. 1578-1585

[20] Long J, Shelhamer E, Darrell T. Fully convolutional networks for semantic segmentation. In: Proceedings of the IEEE Conference on CVPR; 2015

[21] He K, Zhang X, Ren S, Sun J. Deep residual learning for image recognition. In: Proceedings of the IEEE Conference on CVPR. 2016. pp. 770-778

[22] Karim F et al. Lstm fully convolutional networks for time series classification. IEEE Access. 2018;**6**:1662-1669

[23] Hochreiter S, Schmidhuber J. Long short-term memory. Neural Computation. 1997;**9**:1735-1780

[24] Le Guennec A, Malinowski S, Tavenard R. Data augmentation for time series classification using convolutional neural networks. In: In: ECML/PKDD Workshop on AALTD. 2016

[25] Zheng Y, Liu Q, Chen E, et al. Time series classification using multi-channels deep convolutional neural networks. In: Proceedings of the 15th ICWAIM. 2014. pp. 298-310

[26] Zhao B et al. Convolutional neural networks for time series classification. Journal of Systems Engineering and Electronics. 2017;**28**(1):162-169

[27] Hatami N, Gavet Y, Debayle J. Classification of time series images using deep convolutional neural networks. In: Proceedings of ICMV 2017, vol. 10696; International Society for Optics and Photonics. 2018

[28] Chen Y, Keogh E, Hu B, Begum N, Bagnall A, Mueen A, et al. The UCR Time Series Classification Archive. Available from: www.cs.ucr.edu/~eamonn/time_series_data/

[29] Yang J, Nguyen MN, San PP, Li X, Krishnaswamy S. Deep convolutional neural networks on multichannel time series for human activity recognition. In: IJCAI. Vol. 15. 2015. pp. 3995-4001

[30] Rad NM, Kia SM, Zarbo C, van Laarhoven T, Jurman G, Venuti P, et al. Deep learning for automatic stereotypical motor movement detection using wearable sensors in autism spectrum disorders. Signal Processing. 2018;**144**:180-191

[31] Groß W, Lange S, Bödecker J, Blum M. Predicting time series with space-time convolutional and recurrent neural networks. In: Proceedings of the 25th ESANN. 2017. pp. 71-76

[32] Stockwell RG, Mansinha L, Lowe RP. Localization of the complex spectrum: The S transform. IEEE Transactions on Signal Processing. 1996; **44**(4):998-1001

[33] Goodwin MS, Haghighi M, Tang Q, Akcakaya M, Erdogmus D, Intille S. Moving towards a real-time system for automatically recognizing stereotypical motor movements in individuals on the autism spectrum using wireless accelerometry. UBICOMP. 2014;**14**

[34] Ugulino W, Cardador D, Vega K, Velloso E, Milidiú R, Fuks H. Wearable computing: Accelerometers' data classification of body postures and movements. In: Advances in Artificial Intelligence-SBIA. 2012, 2012. pp. 52-61

[35] Ma Y, Genton MG, Parzen E. Asymptotic properties of sample quantiles of discrete distributions. Annals of the Institute of Statistical Mathematics. 2011;**63**:227-243

Section 3

Time Series Forecasting in Real-World Problems

Chapter 5

Forecasting Shrimp and Fish Catch in Chilika Lake over Time Series Analysis

Rohan Kumar Raman and Basanta Kumar Das

Abstract

Chilika lagoon (a Ramsar site) is a large source of fish production and biodiversity situated in the east coast of India, Odisha. Shrimp landings contribute an average of 4185 MT (Metric Ton) around 35% of total fish production. In this study, SARIMA (Seasonal Auto Regressive Integrated Moving Average) model has been developed on quarterly time series shrimp catch data during the year 2001–2015 and forecasted up to 2018. The best model was selected on Akaike Information Criteria (AIC) and Bayesian Information Criterion (SBC). Results showed that maximum average shrimp landings were observed in the first quarter period (summer season), whereas maximum variation in catch was observed during second quarter Q2 (monsoon season) and lowest variation in the fourth quarter Q4 (winter season) catch during the year 2001–2015. The developed time series SARIMA $(0,1,1)(0,1,1)_4$ model was found to be the best fitted model for the shrimp landings in the lagoon. This article also delineates the application of SARIMAX model (SARIMA with regressors) using monthly catch prediction of fisheries in the Chilika Lake. The developed model is validated with less than 10% errors showing increasing fish catch in the upcoming years by maintaining the present lake condition.

Keywords: Chilika, shrimp, fisheries, time series, forecasting

1. Introduction

Coastal lagoon is a junction of fresh, brackish, and marine water system and known for its large fish species production and diversity among aquatic ecosystem [1, 2]. The fresh and saline water combination makes it more liable for attracting different finfish and shellfish species in the aquatic ecosystem. The Chilika lake (lagoon) (a Ramsar site of International importance) is the largest brackish water lagoon in the eastern coast of India situated between latitude 19°20′13.06″–19° 54′47.02″ N to longitude 85°06′49.15″–85°35′32.87″ E in the humid tropical climatic zone in the state of Odisha, India (**Figure 1**). The lagoon is composed of fresh, brackish and marine water ecosystems and divided into outer channel, central sector, northern sector and southern sector. The lagoon is a rich source of fisheries production and biodiversity, which harbours more than 29 species of shrimp from 8 families [3], providing livelihood around 0.2 million people around it. Fisheries resources in Chilika lake account more than 71% total valuation of the lake ecosystem [4]. The lake has been supporting around 200 million rupees to the state

IntechOpen

economy and contributing to the earning of valuable foreign exchange. However, continued natural changes and unabated anthropogenic pressure fisheries suffered the most in terms of both yield and biodiversity in the last two decades [5]. Before restoration (opening of the new lake mouth, year 2000–2001) of the lagoon, the *Penaeid prawns'* population decreases due to the failure in breeding as the lake mouth was shifted far (about 30 Km) from the lake proper, and the confluence point of outer channel (recruitment route) at Magarmukh was silt-choked [5]. Average fisheries catch increased around 520% after four years of new lake mouth opening in 2000. Northern sector was found to have maximum catch percentage of total catch followed by central sector, southern sector and outer channel. The commercial fish catch of the lagoon is composed of 12 fish groups namely mullets, clupeoides, perches, threadfins, crockers, beloniformes, catfishes, tripod fishes, cichlids, murrels, featherbacks and others. After the opening of new lake mouth and desiltation of outer channel, the salinity of the lake increases due to high intake of sea water, resulting favourable for effective recruitment of all economic fishes except cichlids. Shrimp production individually contributed 35% of total fish production in the Chilika lagoon. Some dominating shrimps are *Fenneropenaeus indicus, Penaeus monodon*, *Metapenaeus dobsoni*, *Metapenaeus monoceros,* etc. in the lagoon. The annual landings of shrimp species fluctuated between 2347.78 Metric Ton (MT) and 6413.78 (MT) during last 15 years (2001–2015) in this lagoon after the opening of new lake mouth. Quarter wise shrimp landings fluctuated between 82.55 (MT) and 745.55 (MT) during the period 2001–2015 in the lagoon. To support the prediction of shrimp catch landing in the lagoon, domain specialists need to develop the forecasting models over time series analysis. Time series data are a sequence of data which are collected at successive equally spaced time interval, and it depends on its past value. Time series analysis built stochastic models based on time correlations of collected data. The main objective of time series analysis in respect to the fishery fields is to describe the underlying structure using input data to provide short-term

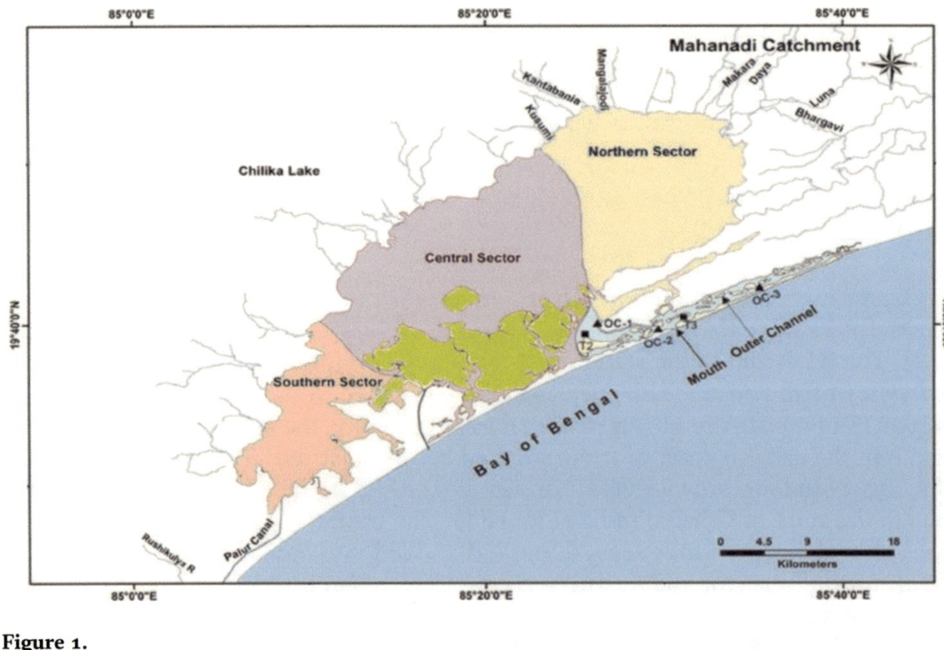

Figure 1.
Map of Chilika lagoon (lake) consisting four ecological sectors (outer channel, central sector, northern sector and southern sector) (Source: [14]).

forecasts of the output variables of a dynamic system [6]. Very well-known Box Jenkin's [7] Auto Regressive Integrated Moving Average (ARIMA) model is used to analyse and forecast the univariate time series data. SARIMA model is composed of ARIMA model including seasonal component of the time series data and very frequently used for time series modelling. Applications of ARIMA and SARIMA models were used to analyse and forecast fisheries data [12–14]; illustration on catch and effort data for the skipjack tuna fleet at Hawaii [8], Saila [9] modelled on monthly average catch per day of rock lobster, Sathianandan and Srinath [10] modelled of all India marine fish landings using ARIMA model, while NunoPrista et al. [11] described the application of SARIMA models for data-poor fisheries with a case study on sciaenid fishery of Portugal. Modelling and forecasting of marine fish production in Odisha, India, using the SARIMA Model [12].

When exogenous variable is included in the ARIMA model, it is known as ARIMAX model. Similarly, SARIMAX model is developed using well-known seasonal ARIMA model with external regressors associated with the time series data. SARIMAX model performs with better accuracy in fish catch prediction than the SARIMA model with minimum error variance [14]. However, the environmental factors' influence as a regressor in the time series ARIMA (SARIMA) model could be quantified by identifying the underlying patterns in periodic time series data using ARIMAX (SARIMAX) model. The influence of water quality parameters of Chilika lake on monthly catch of commercial fisheries (Beloniformes: order) and total fisheries of the lake was modelled using SARIMAX modelling approach [13, 14]. In this case study, SARIMA model for modelling and forecasting the 15 years quarterly time series shrimp catch data of Chilika lagoon of post restoration period since 2001 was developed as case study and further using this model to forecast shrimp catch landings up to 2018. Very few studies have been conducted based on time series model based forecasting of fish species in general and shrimp in particular of Chilika lagoon. The aim of this study is to develop the best fitted time series forecasting model for shrimp landing in Chilika lagoon and further model based catch forecast up to the period 2018. Since catch prediction in advance would be necessary for appropriate planning and designing of the national fishery development plan for sustainable exploitation of fisheries of the said water bodies, this shrimp forecasting study in the Chilika lagoon will be immensely useful in decision making for the policy makers and lagoon managers for sustainable fisheries production and management. In this chapter, a case study of shrimp prediction model using quarterly time series data of the Chilika lagoon was developed, and further prediction model of fisheries and commercial fish (Beloniformes: order) in the Chilika lagoon was also discussed.

2. Materials and methods

2.1 Study area

Chilika Lagoon (**Figure 1**) water spread area fluctuates between monsoon and dry season at maximum of 1165 km^2 to minimum of 906 km^2, respectively [15, 16]. The water quality of lagoon is influenced by the influx of sea water from West Bengal, Mahanadi distributaries and from the western catchment rivers. The lagoon is an assemblage of shallow to very shallow marine, brackish and freshwater ecosystems [15] characterised by the lagoon with marine, brackish and freshwater fisheries. Shrimp landing plays a significant contributor in commercial landings after the opening of new lagoon mouth as a part of hydrological intervention for eco-restoration of the lake in September 2000.

2.2 Sample data collection

The quarter wise estimated total shrimp catch (MT) data and monthly total fish catch (MT) data along with physicochemical parameters of water quality [14] of the Chilika lagoon for the period April 2001 to March 2015 were collected from Chilika Development Authority (CDA). The systematic random sampling methods with landing centre approach [17–19] modified with site-specific conditions followed on monthly basis were used to catch estimation in Chilika lagoon. In this study, quarter wise (seasons) catch of the shrimp was taken; such as first quarter (Q1) consists of March, April and May months together (summer season); second quarter (Q2) consists of June, July and August months (monsoon season); third quarter (Q3) consists of September, October and November months (post monsoon season) and fourth quarter (Q4) consists of December, January and February months (winter season). This quarter wise shrimp catch and total fish catch data for the period 2001 to 2015 of the Chilika lagoon were used for the development of SARIMA time series prediction modelling, and SARIMAX model has been described using monthly total fish catch data with the physicochemical parameters of water quality of the lagoon [14].

2.3 Resource distribution mapping

Annual shrimp catch distribution mapping at 5-year interval is presented using ArcGIS version 9 software.

2.3.1 SARIMA and SARIMAX model development

ARIMA model is developed on stationary data, but very few data sets in the fisheries field are found to be stationary in nature. It becomes mandatory to test the time series data sets for stationarity before using for modelling. Augmented Dickey-Fuller [20, 21] (ADF) test used to test stationary of the original time series data. If data were found to be nonstationary, then it is made stationary by transformation procedure. Stationary time series data were further used for SARIMA model development. SARIMA model was developed by Box-Jenkins [7] Seasonal Auto Regressive Integrated Moving Average (SARIMA) model for seasonal quarterly time series data following four steps as model identification, parameter estimation and model validation (diagnostic checking) and finally forecasting by the following methodology:

SARIMA model defined as ARIMA with seasonal parameters denoted as ARIMA (p,d,q) (P,D,Q) s,

where p = auto regression (AR) order, d = differencing order, q = moving average order, P = seasonal AR order, D = seasonal differencing order, Q = seasonal MA order and s is seasonality using the back shift operator B (the operator B is such that $Bz_t = z_{t-1}$) is expressed as,

$$\varphi(B^s)\phi(B)\nabla^d\nabla_s^D z_t = \Theta(B^s)\theta(B)\varepsilon_t \tag{1}$$

where,

$$\nabla = 1 - B$$
$$\nabla^s = 1 - B^s$$
$$\phi(B) = 1 - \phi_1 B - \cdots - \phi_p B^p$$
$$\theta(B) = 1 - \theta_1 B - \cdots - \theta_q B^q$$
$$\varphi(B^s) = 1 - \varphi_1 B^s - \cdots - \varphi_P (B^s)^P$$
$$\Theta(B^s) = 1 - \Theta_1 B^s - \cdots - \Theta_Q (B^s)^Q$$

Parameters of the model are
$\phi_1, \cdots, \phi_p, \theta_1, \cdots, \theta_q, \varphi_1, \cdots, \varphi_P, \Theta_1, \cdots, \Theta_Q$ and σ^2.
Now, SARIMA model for time series data y_t using Eq. (1) can be defined as;

$$y_t = \varphi(B^s)\phi(B)\nabla^d\nabla_s^D y_t + \Theta(B^s)\theta(B)\varepsilon_t \quad (2)$$

Moreover, the SARIMAX model (seasonal ARIMA with explanatory variable) can be represented as a time-series forecasting model using the multiple regressions with seasonal ARIMA model that takes care of the residual's serial correlations.

Further, SARIMAX model can be defined using Eq. (2) and regressors as follows;

$$y_t = \varphi(B^s)\phi(B)\nabla^d\nabla_s^D y_t + \Theta(B^s)\theta(B)\varepsilon_t + \beta x_t \quad (3)$$

Where, x_t is the input series at time t and β is the regression coefficients of the input series considering the assumptions of ARIMA and regression model together [22].

Here, the sequence $\varepsilon_t^{'}S$ is independently and identically distributed (*i.i.d*) random variable with zero mean and constant variance σ^2, which represents the error term in the model.

Model parameters estimation and its significance were performed using statistical software such as SAS, R, MATLAB, SPSS, etc. Generally 70% data sets were taken for model development and 30% data for model validation for long time series data, but one can change the percentage based on the data availability. Model development approach (a) model identification: model parameters are identified by investigating autocorrelation function (ACF) and partial auto correlation function (PACF) of the time series data (see [23] for more details); (b) parameter estimation: identified model parameters are estimated using various recursive statistical algorithms computation [23]; (c) model validation: inspection of residuals for no significant lags using ACF. The best model was selected based on model selection criteria and finally forecast is done data using best model for the given period of time. The known model selection criteria (minimum is better), such as Akaike [24], known as Akaike Information criterion (AIC), Bayesian Information Criterion proposed by Schwartz [25], known as SBC and R square fit statistics, RMSE (root mean square error), are used for identification of the best fitted model. The minimum is preferable criteria except R square (maximum) for identifying best selection model. Finally, the Ljung and Box [26] statistics Q based on the autocorrelations of the residuals were used for testing adequacy of estimated models. Finally, forecasting or prediction for the short term period is done based on best fitted model. For shrimp catch forecasting modelling, Quarter wise shrimp catch data for the period April, 2001 to March, 2013 was used for best model selection and estimation and the data April, 2014 and March, 2015 were used for validation of the model for Chilika lagoon. Further based on developed model, quarterly shrimp catch was forecasted up to 2018. Further SARIMAX model is described using monthly total fish catch data with the physicochemical parameters of water quality as a factor of regressor of the lagoon as regressors [14]. The total fish data sets for the period April 2001-- March 2011 were used for training data sets, and the period April 2011–March 2014 was used as testing and validation data sets for the model.

3. Results

The quarter wise shrimp catch landings (total, mean and standard deviation) of Chilika lagoon are shown in **Table 1**. The total catch in first quarter Q1 (7427.36)

Time (quarter)	Total catch (MT)	Mean catch (MT)	Std. (MT)
Q1	7427.36	495.15	126.61
Q2	7033.78	468.91	164.69
Q3	3216.19	214.41	84.03
Q4	3226.56	215.10	55.60

Table 1.
Quarter wise total, mean and standard deviation of shrimps landing in Chilika lagoon for the period 2001–2015.

was found to be maximum followed by second quarter Q2 landings (7033.78). The catch of third quarter Q3 (3216.19) and fourth quarter Q4 (3226.56) are nearly equal for the period 2001–2015. The quarterly average landings Q1 (495.15) are found to be more than Q2 (468.91), but the variation in catch in Q2 (164.69) is higher than Q1 (126.61). The mean catch of Q3 (214.41) is similar to that of Q4 (215.10), but the variation in catch in Q3 (84.03) is higher than Q4 (55.60). The variation in quarter wise shrimp catch data in Chilika lagoon was shown in **Figure 2**. Temporal variation using image mapping of the lagoon in **Figure 3** showed that catch increases in the first decade but decreases thereafter, and in 2015, catch was found comparatively low with respect to 2010–2012. The highest peak in catch was observed during 2011 and then starts decreasing till 2015 (**Figure 3**).

The best fitted SARIMA model was selected based on model selection criteria (AIC and SBC), and SARIMA $(0,1,1)(0,1,1)_4$ model with lowest AIC (509.23) and SBC (517.26) found the best fitted model **Table 2**. The parameter's estimate of best-fitted SARIMA $(0,1,1)(0,1,1)_4$ model in **Table 3** showed nonseasonal moving average (component) parameter Q at lag 1 (0.71, $p < 0.001$) and seasonal moving average parameter (0.78, $p < 0.001$) at lag 4 as significant effect in the developed model. Two catches of second quarter for the year 2005 and 2011 showed the significant outliers, and catch showed significantly increase during this period. The R^2 value 0.70 showed the good fit of the SARIMA $(0,1,1)(0,1,1)_4$ model. The residual of the developed SARIMA model was tested by Ljung, and Box test ($p > 0.05$) showed the adequacy of the fitting of the estimated models. The developed SARIMA model was validated with the actual shrimp catch data (**Table 4**). The annual catch prediction error for the developed model was found below 10%,

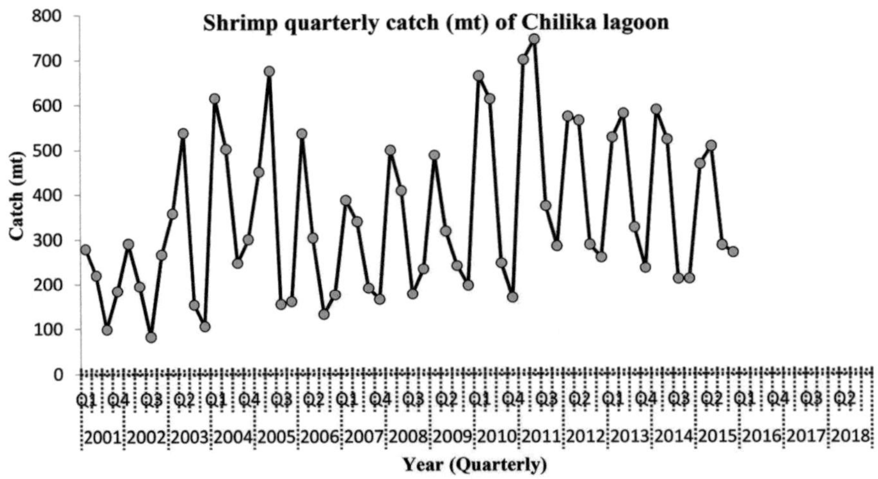

Figure 2.
Quarter wise shrimp catch (MT) in Chilika lagoon for the period 2001 to 2015.

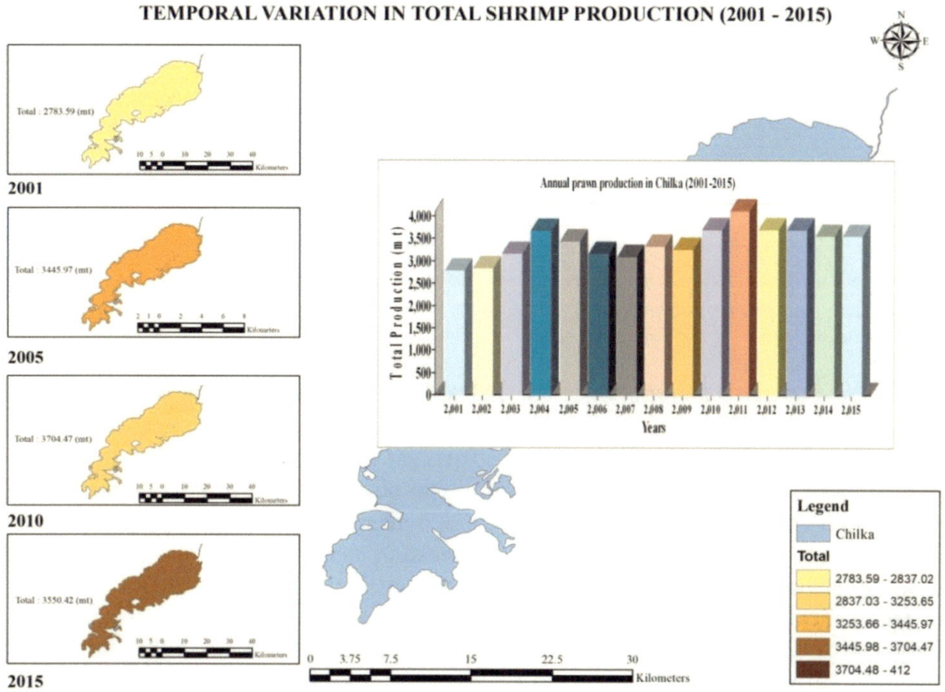

Figure 3.
Temporal variation of shrimp production of Chilika lagoon ecosystem from 2001 to 2015.

Sl. No.	SARIMA model	AIC	SBC
1	SARIMA$(0,1,1)(0,1,1)_4$	509.23	517.26
2	SARIMA$(0,0,1)(1,1,1)_4$	519.19	529.32
3	SARIMA$(1,0,1)(1,1,1)_4$	520.88	533.03
4	SARIMA$(1,0,1)(1,0,1)_4$	558.60	573.26
5	SARIMA$(1,0,0)(1,1,1)_4$	518.55	528.10
6	SARIMA$(1,1,1)(1,0,0)_4$	585.03	595.85
7	SARIMA$(0,0,1)(0,1,1)_4$	517.35	525.46
8	SARIMA$(2,0,0)(1,1,1)_4$	519.94	532.10

Table 2.
Comparisons of best fitted SARIMA model with fit statistics (AIC, SBC) for quarter wise shrimp landings in Chilika lagoon for the period 2001–2015.

Model parameter	Estimate	Std. Error	Prob>|T	Fit Statistics
Q (Lag 1)	0.71	0.11	<.0001	AIC = 509.23
q (Lag 4)	0.78	0.14	<.0001	SBC = 517.26
2011:Q2 (catch)	205.16	82.48	0.01	R^2 = 0.70
2005:Q2 (catch)	266.90	82.31	0.002	

Table 3.
Best fitted SARIMA $(0,1,1)(0,1,1)_4$ model parameter's estimate and their significance for quarter wise shrimp landings in Chilika lagoon for the period 2001–2015.

Year	Actual shrimp catch (MT)	Predicted shrimp catch (MT)	% error in catch prediction
2014	1536.418	1676.94	9.14
2015	1535.96	1520.201	1.02
2016		1605.134	4.5
2017		1627.088	5.9
2018		1649.041	7.36

Table 4.
Yearly actual versus predicted shrimp catch forecast with % prediction error using SARIMA $(0,1,1)(0,1,1)_4$ model in Chilika lagoon for the year 2014–2018.

e.g., for year 2014 (9.14%) and 2015 (1.02%). The percentage forecast error in shrimp catch for the year 2016, 2017, and 2018 was found to be 4.5, 5.9, and 7.36%, respectively (**Table 4**). The quarter wise forecast of shrimp catch for the period 2016–2018 with 95% upper and lower catch was shown in **Table 5**. The quarter wise actual catch versus predicted catch with 95% confidence limit of shrimp catch for the period of 2001 to 2018, showing an increase in catch with respect to 2015, is shown in **Figure 4**.

An application of SARIMAX model for total fish production forecasting is analysed for Chilika lagoon [14]. Mullets, clupeids, engraulids, perches, catfishes, sciaenids, threadfins, cichlids, tripod fishes, featherbacks, murrels, carps and

Year (quarterly)	Actual catch (MT)	Forecasted catch (MT)	Upper 95% catch (MT)	Lower 95% catch (MT)
2014(Q1)	589.113	604.435	795.914	412.956
2014(Q2)	522.827	546.637	738.091	355.183
2014(Q3)	211.991	310.37	501.812	118.929
2014(Q4)	212.487	215.502	436.937	94.0664
2015(Q1)	468.227	567.049	758.48	375.618
2015(Q2)	508.377	482.476	673.891	291.06
2015(Q3)	287.953	243.258	434.666	51.8502
2015(Q4)	271.403	227.418	418.822	36.0139
2016(Q1)		555.174	746.527	363.821
2016(Q2)		519.439	718.554	320.324
2016(Q3)		278.325	484.91	71.7396
2016(Q4)		252.196	465.991	38.4016
2017(Q1)		560.662	794.815	326.509
2017(Q2)		524.927	768.392	281.463
2017(Q3)		283.813	536.246	31.3805
2017(Q4)		257.685	518.778	93.4083
2018(Q1)		566.151	848.368	283.933
2018(Q2)		530.416	823.301	237.531
2018(Q3)		289.302	592.479	183.875
2018(Q4)		263.173	576.304	149.957

Table 5.
Quarter wise actual versus forecasted shrimps catch with upper 95% catch (MT) and lower 95% catch (MT) prediction error using SARIMA modelling approach for the period 2014 (Q1) to March 2018 (Q4) of Chilika lagoon.

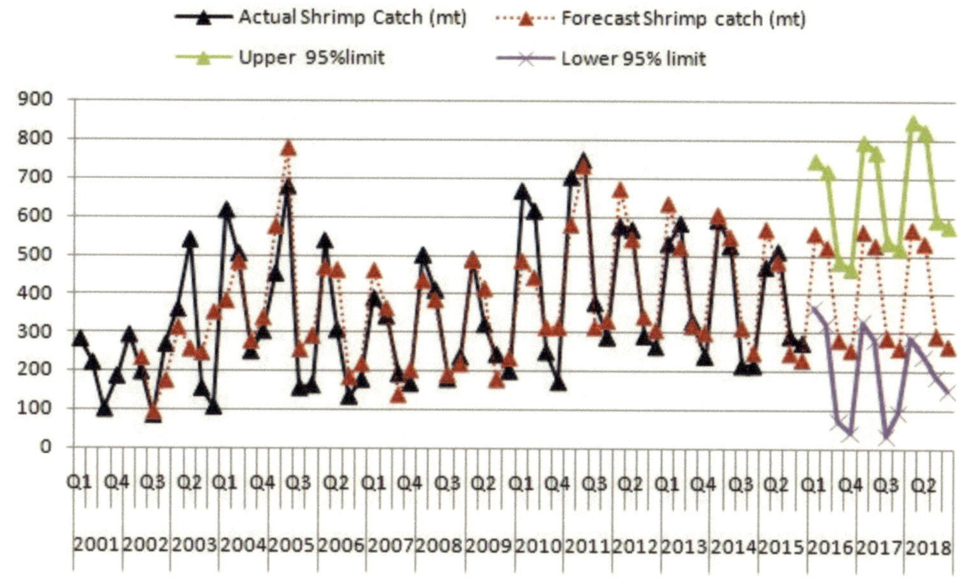

Figure 4.
Quarter wise actual versus predicted shrimp catch (MT) with 95% confidence limit of Chilika lagoon for the period 2001 to 2018.

minnows with some miscellaneous catch are the major fish catch of the lagoon with an average annual estimate of fish landing 12,000 tons per annum from 2000–2001 to 2014–2015. The fish landing was fluctuated between 10,286.34 tone (2003–2004) and 6463.92 tone (2006–2007) with average at 7725.42 tone contributing 63.65% of total fish landing of the whole lagoon. Around 64.88% of commercial catch was contributed by the yield of fish since last 15 years of post-restoration period. More than 90% fish species which contributed to the commercial landing of Chilika are of brackish or marine habitats. SARIMAX model was developed using monthly time series fish catch data, and regressor is used as a factor (combination of lagoon

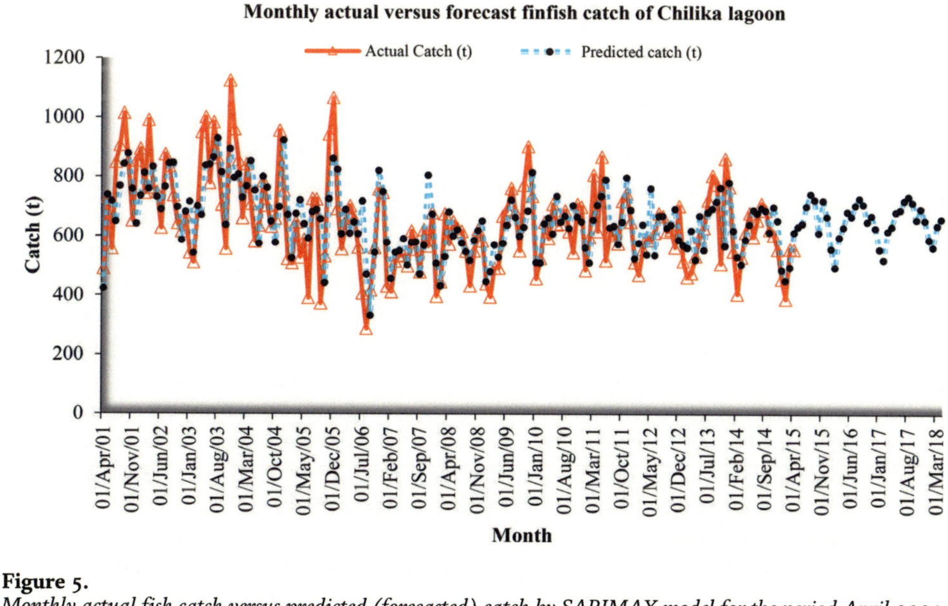

Figure 5.
Monthly actual fish catch versus predicted (forecasted) catch by SARIMAX model for the period April 2001 to March, 2018 of Chilika lagoon (Source: [14]).

temperature and salinity for the same period of April 2000–2001 to March 2014–2015 of the Chilika lagoon). The data sets for the period April 2001–March 2011 were used for training sets and April 2011 to March 2015 were taken as validation of the model. SARIMAX (1,0,0)(2,0,0)12 was found to be the best fitted model. Here, SARIMA (1,0,0)(2,0,0)12 model is also compared with the SARIMAX (1,0,0)(2,0,0)12 (SARIMA (1,0,0)(2,0,0)12 with regressors as factor) and found that the R square value is greater for SARIMAX than SARIMA and root mean square error is less for SARIMAX than the SARIMA model, which reveals that SARIMAX model is better performer than the SARIMA model for the forecasting of the total fish catch of the lagoon. The regressor, i.e., temperature and salinity used for SARIMAX development, showed significantly positive influence ($p < 0.05$) on the fish catch in the lagoon. Using SARIMAX model, forecasting is done for the fish production within the error catch less than 10% (for details please see [14]; **Figure 5**). Total fish catch forecasting also showed an increasing catch in the upcoming year till 2018 with respect to the base year 2015.

4. Discussion

In Chilika lagoon, fisheries in general and shrimp catch in particular play an important role in support of livelihoods of fishermen around it. The quarterly catch variation in shrimp landings in lagoon could be due to direct or indirect influence by hydrology and environmental condition of the lagoon system [27–29]. Average shrimp landings during first quarter Q1 were maximum, which reflects the summer season (March, April and May) catch and the lowest in third quarter Q3 representing the post-monsoon season (September, October and November). The maximum variation in shrimp catch was observed during second quarter Q2 representing monsoon season and lowest variation in fourth quarter Q4 winter season (December, January and February). Seasonal and environmental variation influences the special trends of fisheries in the estuaries system [30]. Physicochemical parameters such as salinity and water temperature influence the distribution of various Penaeid shrimp [31]. The physicochemical parameters of the Chilika lagoon are very much influenced by the seasonal variation as it is connected with the fresh water river by one side and the Bay of Bengal on the other side. The fresh water flow showed positive, negative and inconsistent influence on fish production in the coastal lagoon system for different fish species across different countries [32]. The major commercial fishes in the lake are mostly migrant species (between sea and lake) and account more than 86% of the total fish diversity [33]. For feeding and breeding purpose, the migrant species enter the lake from the sea. The lagoon environment becomes favourable in post winter through summer due to the more availability of food. The higher temperature and salinity phase triggers spawning activity of many clupeid species which contribute more than 26% of total annual catch. Hence, in general, more number of fish species abundance is influenced by the temperature and salinity of the lagoon. As reported, the total fish production showed an increase trend in catch with the increase in temperature and salinity [14]. Similar observation was made as marine-dependent species increased from February month to onset of monsoon (June) [19] in Chilika lagoon. Moreover, during monsoon season, the nutrient inflow in the lake comes through catchment runoff and fresh water, which induces plankton production during post winter and summer (high salinity and temperature) period.

The information on fisheries catch prediction in advance plays a crucial role for managing fisheries resources in the lagoon ecosystem. In fisheries, several time series models were developed and forecasted for fisheries catch data such as ARIMA model, multiple linear regression, non-linear regression and dynamic models, Gaussian autoregressive model, etc. [34–36], but in general, ARIMA models have been widely used for better forecasting. SARIMA model is found good fit for short time series catch data and used as common forecasting model by several researchers [34, 37, 38]. This study also analysed and forecast the quarterly shrimp landing of Chilika lagoon using SARIMA time series modelling approach using data for the period 2001–2015. The best fitted SARIMA $(0,1,1)(0,1,1)_4$ model developed for the shrimp catch in Chilika lagoon was validated with actual quarterly catch data for the period 2014–2015 and further quarter wise forecast of the shrimp catch up to the year 2018. The forecast of shrimp catch for the years 2016, 2017, and 2018 shows an increase in catch by 4.5, 5.9, and 7.36% with respect to the base year 2015 if the environmental condition of the lagoon remains the same. SARIMAX (1,0,0)(2,0,0) 12 also showed the total fish increasing catch up to year 2018 with respect to the base year 2015. These model-based fish catch forecasting for Chilika lagoon results will be useful for understanding fish catch forecast with acceptable accuracy that will enable lagoon fish managers to facilitate fisheries management by predicting the lagoon fisheries progress toward fish production. Reliable fish catch production forecasting methods are important for managers for confident decision making in fishery management. Hence, the study suggests the regular monitoring and maintenance of fish catch and water quality data to be continued for assessing fisheries trends to enable corrective steps for short- and long-term management interventions in the lagoon.

5. Conclusions and future work

Time series SARIMA $(0,1,1)(0,1,1)_4$ model and SARIMAX (1,0,0)(2,0,0)12 model are statistical forecasting models for predicting time series quarterly catch of shrimp production and monthly total fish production in Chilika lagoon. Seasons play a significant influence in shrimp and total fish production in the lagoon. Model prediction showed shrimp and total fish catch of the lagoon will increase in the upcoming years by maintaining the present environmental and ecological conditions of the lagoon. Summer and monsoon seasons have explicit influence on shrimp production in the lagoon. Salinity and temperature of the lagoon positively influence on the total fish catch in the lagoon. Model predicts that shrimp and total fish catch will increase in the upcoming years till 2018 with respect to the base year 2015 in the lagoon. Therefore, continuous monitoring in the lagoon is essential to manage its rich and productive fishery resources, as well as its ecological integrity. This information would provide kind support to the managers of Chilika Development Authority to continue the present efforts for getting a good rich production for the livelihood of the fishermen dependent upon and sustainable management of fisheries. This time series model based study will be useful for forecasting fish production in different lagoon or aquatic system. Besides the catch and water quality monitoring, some related information regarding average monthly influx of tides in respect of amplitude and duration may be recorded for correlation with the catch variability of important species and consideration in trend analysis.

Acknowledgements

The authors like to acknowledge the Chilika Development Authority (CDA), Bhubaneswar, Odisha, India, for providing time series fish catch data of Chilika lagoon for the present study. We are thankful to ICAR-CIFRI, Barrackpore,

Kolkata, India, for providing laboratory facilities and other logistics for completion of this research work.

Author details

Rohan Kumar Raman and Basanta Kumar Das*
ICAR-Central Inland Fisheries Research Institute, Barrackpore, Kolkata, India

*Address all correspondence to: basantakumard@gmail.com

References

[1] Elliot M, Hemingway KL. Fishes in Estuaries. 2nd ed. Oxford: Blackwell Science; 2002. p. 636

[2] George B, Kumar JIN, Rita NK. Study on the influence of hydro-chemical parameters on phytoplankton distribution along Tapi estuarine area of Gulf of Khambhat, India. Egyptian Journal of Aquatic Research. 2012;**38**: 157-170

[3] Mohapatra A, Mohanty SK, Mishra SS. Fish and Shellfish Fauna of Chilika Lagoon: An updated checklist. In: Venkataraman K, Sivaperuman C, editors. Marine Faunal Diversity in India. Elsevier Inc. Publications; 2015. pp. 195-224. DOI: 10.1016/B978-0-12-801948-1.00013-6

[4] Ritesh K. Economic valuation of Chilika lagoon. Chilika Newsletter. 2003;**4**:24-26

[5] Mohanty SK, Bhatta KS, Mohanty RK, Mishra S, Mohapatra A, Pattnaik AK. Eco-Restoration Impact on Fishery Biodiversity and Population Structure in Chilika Lake. Lakes and Coastal Wetlands Conservation, Restoration and Management. New Delhi: Capital Publishing Company; 2007. ISBN-81-85589-51-0

[6] Rothschild BJ, Smith SG, Li H. The application of time series analysis to fisheries populations assessment and modelling Stock Assessment: Quantitative Methods and Applications for Small-Scale Fisheries, Florida: CRC Press; 1996. pp. 354-402

[7] Box GEP, Jenkins GM. Time Series Analysis: Forecasting and Control. San Francisco: Holden-Day; 1976

[8] Roy M. Using Box-Jenkins models to forecast fishery dynamics: Identification, estimation and checking. Fishery Bulletin. 1981;**78**(4):887-896

[9] Saila SB, Wighbout M, Lermit RJ. Comparison of some time series models for the analysis of fisheries data. Journal du Conseil. 1980;**39**:44-52

[10] Sathianandan TV, Srinath M. Time series analysis of marine fish landings in India. Journal of the Marine Biological Association of India. 1995;**37**:171-178

[11] Prista N, Diawara N, Costa MJ, Jones C. Use of SARIMA models to assess data-poor fisheries: A case study with a sciaenid fishery of Portugal. Fishery Bulletin. 2011;**109**:170-185

[12] Raman RK, Sathianandan TV, Sharma AP, Mohanty BP. Modelling and forecasting marine fish production in Odisha using seasonal ARIMA model. National Academy Science Letters. 2017;**40**(6):393-397

[13] Raman RK, Sahoo AK, Mohanty SK, Das BK. Forecasting of commercial fish (Beloniformes: Order) catch in Chilika lagoon, Odisha, India. Journal of the Inland Fisheries Society of India. 2017; **49**(1):55-63

[14] Raman RK, Mohanty SK, Bhatta KS, Karna SK, Sahoo AK, Mohanty BP, et al. Time series forecasting model for fisheries in Chilika lagoon (a Ramsar site, 1981), Odisha, India: A case study. Wetlands Ecology and Management. 2018;**26**(4):677-687

[15] World Bank. Scenario assessment of provision of environmental flows to Lake Chilika from Naraj Barrage, Orissa, India. Reports from the environmental flows window of the bank Netherlands water partnership programme (World Bank) to the Government of Orissa, India; 2005. p. 40

[16] Mohapatra ARK, Mohanty SK, Mohanty, Bhatta KS, Das NR. Fisheries enhancement and biodiversity

assessment of fish, prawn and mudcrab in Chilika lagoon through hydrological intervention. Wetlands Ecology and Management. 2007;**15**(3):229-251

[17] Biradar RS. Fisheries Statistics Course Manual No-14. Bombay: Central Institute of Fisheries Education (ICAR); 1988. p. 229

[18] Gupta RA, Mandal SK, Paul S. Methodology for collection and estimation of inland fisheries statistics in India. Central Inland Capture Fisheries Research Institute (ICAR), Barrackpore, West Bengal Bull. No. 58 (Revised); 1991. p. 64

[19] Jhingran VG, Natarajan AV. Study of the fishery and fish populations of the Chilika lake during the period 1957–65. Journal of Inland Fisheries Society of India. 1969;**1**:47-126

[20] Dickey DA, Fuller WA. Distribution of the estimators for autoregressive time series with a unit root. Journal of the American Statistical Association. 1979; **74**:427-431

[21] Dickey DA, Fuller WA. Likelihood ratio statistics for autoregressive time series with a unit root. Econometrica. 1981;**49**:1057-1072

[22] Andrews B, Dean M, Swain R, Cole C. Building ARIMA and Arimax Models for Predicting Long-Term Disability Benefit Application Rates in the Public/Private Sectors. Society of Actuaries; 2013

[23] Cryer JD. Tome Series Analysis. Boston: Duxbury Press; 1986

[24] Akaike H. Use of an information theoretic quantity for statistical model identification. In: Proceedings of the 5th Hawaii International Conference on System Sciences; 1972. pp. 249-250

[25] Schwartz G. Estimating the dimension of a model. The Annals of Statistics. 1978;**6**:461-464

[26] L-Jung GM, Box GEP. On a measure of lack of fit in time series models. Biometrika. 1978;**65**:297-303

[27] Blaber SJM, Blaber TG. Factors affecting the distribution of juvenile estuarine and inshore fish. Journal of Fish Biology. 1980;**17**:143-162

[28] Elliott M, O'Reilly MG, Taylor CJL. The forth estuary: A nursery and overwintering area for North Sea fishes. Hydrobiologia. 1990;**195**:89-103

[29] Morin B, Hudon C, Whoriskey FG. Environmental influences on seasonal distribution of coastal and estuarine fish assemblages at Wemindji, eastern James Bay. Environmental Biology of Fishes. 1992;**35**:219-229

[30] Pomfret JR, Elliot M, Reilly MGO, Philips S. Special and temporal patterns in the fish communities in the U. K. North Sea estuaries. In: Elliot M, Ducrotoy JP, editors. Estuaries and Coast: Special and Temporal Intercomparisons, 2077–284. Fredenborg: Olsen and olsen; 1991

[31] Walsh CJ, Mitchell BD. Factors associated with variations in the abundance of epifaunal caridian shrimps between and within seagrass meadows. Marine and Freshwater Research. 1998;**49**:769-777

[32] Gillson J. Freshwater flow and fisheries production in estuarine and coastal systems: Where a drop of rain is not lost. Reviews in Fisheries Science. 2011;**19**(3):168-186. DOI: 10.1080/10641262.2011.560690

[33] Mohanty SK, Mishra SS, Khan M, Mohanty RK, Mohapatra A, Pattnaik AK. Ichthyofaunal diversity of Chilika Lake, Odisha, India: An inventory, assessment of biodiversity status and comprehensive systematic checklist (1916–2014). Check List. 2015;**11**(6): 1-19

[34] Stergiou KI. Describing and forecasting the sardine-anchovy complex in the eastern Mediterranean using vector autoregressions. Fisheries Research. 1991;**11**:127-141

[35] Stergiou KI, Chritou ED, Petrakis G. Modelling and forecasting monthly fisheries catches: Comparison of regression, univariate and multivariate time series methods. Fisheries Research. 1997;**29**:55-95

[36] Romilly P. Time series modelling of global mean temperature for managerial decision-making. Journal of Environmental Management. 2005;**76**: 61-70

[37] Pajuelo JG, Lorenzo JM. Analysis and forecasting of the demersal fishery of the Canary Islands using an ARIMA model. Scientia Marina. 1995;**59**:155-164

[38] Hae-Hoon P. Analysis and prediction of walleye Pollock (*Theragra chalcogramma*) landings in Korea by time series analysis. Fisheries Research. 1998;**38**(1):1-7

Chapter 6

Using Gray-Markov Model and Time Series Model to Predict Foreign Direct Investment Trend for Supporting China's Economic Development

Yanyan Zheng, Tong Shu, Shou Chen and Kin Keung Lai

Abstract

Foreign direct investment (FDI) is one of the important factors affecting China's economic development, the prediction of which is the basis of its development and decision-making. Based on elaborating the significant role in China's economic growth and the status quo of utilizing foreign investment over the period between 2000 and 2016, this chapter attempts to construct Gray-Markov model (GMM) and time series model (TSM) to forecast the trend of China's utilization of FDI and then compares the precision of two different prediction models to obtain a better one. Results indicate that although it is qualified, traditional Gray model needs to be optimized; GMM is built to help modify the result, improve Gray-related degrees, and narrow the gap with real value. Comparing the accuracy of GMM with that of TSM, we can conclude that the fitting effect of GMM is better. To increase the credibility of these results, this chapter is based on the data of Beijing and Chongqing from 1990 till 2016, also verifying that the fitting effect of GMM is superior to that of the TSM. Then, we can safely draw a conclusion that the prediction model of GMM is more credible, which has a certain referencing value for the utilization of FDI.

Keywords: foreign direct investment (FDI), Gray-Markov model (GMM), time series model (TSM)

1. Introduction

In the light of the definition of the International Monetary Fund (IMF) and the Organization for Economic Cooperation and Development, foreign direct investment (FDI) is an investment in the form of a controlling ownership in a business in one country by an entity based in another country. The primary purpose of the host country in attracting FDI is to promote the country's economic development and industrial upgrading. This will facilitate domestic enterprises to improve their technology and quality, gradually supporting the development of foreign enterprises to enter the global value chain [1]. Influencing the supply chain system, FDI has significantly promoted the sound and rapid development of the national economy.

IntechOpen

Therefore, it is necessary to focus on the future tendency of FDI in the supply chain system when we investigate the transformation and innovation of Chinese economy.

Since the late 1970s, FDI attracted by China has been steadily increasing, regardless of the changes and fluctuation of the international economic environment and the total flow of FDI globally. Statistically, over the period from 1979 to 2010, China's actual use of FDI amounted to $1048.31 billion [2], and FDI keeps a rapid growth. According to the data of Ministry of Commerce of the People's Republic of China (PRC) (**Figure 1**), the FDI in China presented a rising trend over the period from 1990 to 2016. The vital roles in the economic development of China are as follows. Firstly, the proportion of basic industries in China declines generally, and the proportion of agricultural output drops by 18% over the period between 1978 and 2011 [3]. Secondly, for a long time, FDI mainly concentrates in secondary and tertiary industries, accelerating the restructuring and upgrading of China's industries [4]. Finally, FDI provides investment capital and promotes the rapid development of China's import and export trade, improving China's status in international trade.

Due to the remarkable role of FDI, a multitude of scholars began to track and study the FDI in developing countries, build analytical framework, and launch a new field of research of FDI in developing countries. The statistics shows that China has become an emerging market for FDI. Dees indicates that FDI has positive effects on the GDP, technological progress, and the improvement of management system [5]. Nourzad considers that FDI promotes economy development through technology transfer [6], while Mah argues that the latter one promotes the former one [7]. Taking the reform policy (implemented in July 2005) as the boundary, Pan and Song explore the impact of the effective exchange rate of RMB on FDI [8]. Research shows that they are in a long-term equilibrium relationship before implementing reform policy. After the policy, the exchange rate of RMB has the Granger causality for FDI, and the appreciation of RMB can promote the flow of FDI. Additionally, De Mello shows that FDI can increase the added value associated with it [9]. Based on the data from 1971 to 2012, Dreher et al. conclude that the membership in international organizations is an essential and decisive factor of FDI

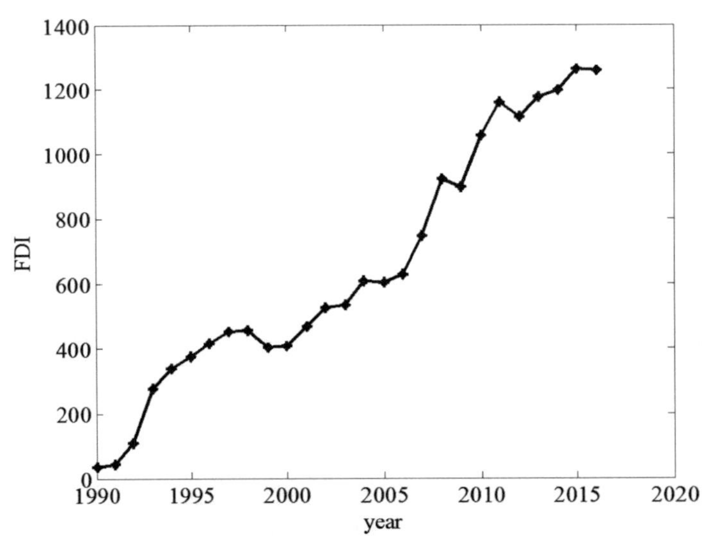

Figure 1.
The horizontal curve of FDI in China.

liquidity and has a promoting effect on FDI mobility [10]. Badr and Ayed do a quantitative study of the relationship between FDI and economic development in South American countries, and they find that FDI can be determined by some economic factors, having no important effect on economic development [11]. Kathuria et al. apply panel data to examining the effectiveness of public policy in attracting FDI [12]. Lin et al. divide the FDI company into five strategies [13]; Brülhat and Schmidheiny estimate the rivalness of state-level inward FDI [14].

The trend of FDI in the future is an important reference for China's economic development. However, much literature focuses on the development of FDI itself and its influencing factors, and there is little research on the future development. This is what we do in this chapter. Currently, the predictive analysis model for economic and trade development can be divided into linear prediction method and nonlinear prediction one. The linear prediction method mainly includes historical average level prediction method, time series prediction method, and Kalman filter prediction method, to name just a few. The nonlinear prediction methods concern Gray theory, Markov chain, support vector machine, and boom prediction method. The historical average prediction algorithm is simple and easy to understand and the parameters can be estimated by using the least squares method. However, it is too simple to accurately reflect the randomness and nonlinearity, and therefore it cannot be applied to unexpected events. The Kalman filter uses the flexible recursive state space model, with the advantages of linear, unbiased, and minimum mean variance. Nevertheless, because the Kalman filter prediction model belongs to the linear model, its performance becomes worse in the nonlinearity and uncertainty [15]. The time series model is simple in modeling, with high prediction accuracy in the case of full historical data. The Gray model can be modeled with less information, handling data easily and having higher accuracy, which can be extensively used in several fields [15–18]. However, Gray model becomes less attractive for time series with large stochastic fluctuation. Markov stochastic process predicts the development and changes of dynamic system according to the transfer probability of different states, and the transfer probability reflects the influence degrees of various stochastic factors and the internal law of the transition states. Therefore, it is more suitable to predict the problems with large stochastic fluctuation. What cannot be ignored is that Markov model requires data to meet the characteristics of no effect. Consequently, when using a simple model, it is very difficult to obtain a better prediction result, and the combination method becomes a popular method.

Through the vector autoregressive moving average (VARMA), Bhattacharya et al. compare and analyze the consumer price index sequence (CPI) and improve the forecasting accuracy [16]. The Gray model (proposed by a Chinese scholar, Professor Deng) and the Markov model (proposed by a Russian mathematician, Markov) have been combined very early, which is called Gray-Markov model (GMM). Based on the Gray prediction model, GMM is used to solve the inaccurate problems resulting from the large random fluctuation of the data and widely promoted in the fields of financial economy, agricultural economy, and resource and energy [17–20]. On the basis of GM(1,1), Li et al. propose an improved GM(2,1) model [21]. Based on the model of GM(1,1) and Markov stochastic process and combining Taylor formula approximation method, Li et al. construct a model of T-MC-RGM(1,1) and verify its validity by the example of thermal power station in Japan [22].

The level of FDI in China is influenced by many factors such as fixed investment, laws and regulations, corporate culture, innovation ability, and financial market stability, among others. To clearly recognize and describe the role of FDI, the foreign investment system is abstracted as a Gray system with no physical prototype and incomplete information, which can be predicted with GM(1,1) model. Meanwhile, the FDI level in the previous year has no direct influence on that in the next year, in line with the no-effect characteristic of Markov stochastic process. On the basis of the previous study of Gray-Markov model, it is used to predict the tendency of FDI in China, addressing the shortcomings of the Gray model for the low precision of the data sample with large fluctuation and compensating for the limitation that the Markov model requires the data to have a smooth process. As a comparison, the time series prediction model is introduced to evaluate FDI. Then, the fitting results are compared to decide the optimal prediction model.

2. Gray-Markov model

Gray-Markov model is a forecasting method integrating the Gray theory with the Markov theory [17–25]. Firstly, GM(1,1) is constructed to obtain the predicted residual value. Then, the error state can be divided according to the residual values, and the error state can be obtained in light of the Markov prediction model. Then, based on the error state and transition matrix, the predicted sequence from GM(1,1) can be adjusted to obtain more precise predicting internals. The traditional GM(1,1) has its advantage in short-term prediction, while it has a poor fitting effect in forecasting the long-range and fluctuating data series. And the benefit of Markov stochastic process is the prediction of the large data series with random volatility. GMM has been proposed by He to predict the yield of cocoon and oil tea in Zhejiang Province. Subsequently, this model is widely used in the prediction of transportation, air accidents, and rainfall. Accordingly, we use GMM to predict FDI of China [26–28].

2.1 Gray model

The Gray system theory, founded and developed by Chinese scholar Deng, extends the viewpoints and methods of general system theory, information theory, and cybernetics to the abstract system of society, economy, and ecology, incorporating the development of mathematical methods to develop the theory and method of Gray system. The modeling process is as follows.

(1) Raw series are

$$X^{(0)} = \left\{x^{(0)}(1), x^{(0)}(2), \ldots\ldots, x^{(0)}(m)\right\} \tag{1}$$

(2) To weaken the randomness of the original data, the accumulated generating series is derived:

$$X^{(1)}(k) = \sum_{i=1}^{k} x^{(0)}(i). \tag{2}$$

(3) Based on the sequence of $X^{(1)}_{(t)}$, a new sequence $Z^{(1)}_{(t)}$ is derived as follows:

$$Z^{(1)}(k) = \frac{1}{2}x^{(1)}(k) + \frac{1}{2}x^{(1)}(k-1) \tag{3}$$

(4) Then, whitened differential equation is obtained:

$$x^{(0)}(k) + aZ^{(1)}(k) = b \tag{4}$$

In Eq. (4) a is development coefficient, b is the parameter of Gray action, and Φ is identification parameter vector. Then, the least squares estimation of parameters satisfies the following equation:

$$\hat{\Phi} = \begin{pmatrix} \hat{a} \\ \hat{b} \end{pmatrix} = \left(B^T B\right)^{-1} B^T Y \tag{5}$$

and

$$B = \begin{pmatrix} -Z^{(1)}(2) & 1 \\ -Z^{(1)}(3) & 1 \\ \vdots & \vdots \\ -Z^{(1)}(m) & 1 \end{pmatrix}, Y = \begin{pmatrix} x^{(0)}(2) \\ x^{(0)}(3) \\ \vdots \\ x^{(0)}(m) \end{pmatrix} \tag{6}$$

By differentiating $x^{(1)}(k)$, a whitened differential equation can be written as $\frac{dx^{(1)}}{dt} + ax^{(1)}(k) = b$

(5) The whitened time response is as follows:

$$\hat{x}^{(1)}(k+1) = \left(x^{(1)}(1) - \frac{\hat{b}}{\hat{a}}\right)e^{(-\hat{a}k)} + \frac{\hat{b}}{\hat{a}} \tag{7}$$

Reducing the sequence of $\hat{x}^{(1)}(k+1)$ $(k = 1, 2, ..., m-1)$, the following sequence is obtained:

$$\hat{X}^{(0)} = \left\{\hat{x}^{(0)}(1), \hat{x}^{(0)}(2), ..., \hat{x}^{(0)}(m)\right\} \tag{8}$$

(6) Model testing

Model test is divided into residual test and Gray-relating test. Residual test is to obtain the difference between predicting value and the actual value. Firstly, the absolute residuals and relative residuals about $X^{(0)}$ and $\hat{X}^{(0)}$ are calculated:

$$\Delta^{(0)}(i) = \hat{x}^{(0)}(i) - x^{(0)}(i)(i = 1, 2, ..., n) \tag{9}$$

$$\phi(i) = \frac{\Delta^{(0)}(i)}{\hat{x}^{(0)}(i)}(i = 1, 2, ..., n) \tag{10}$$

Then, below is the average value of relative residuals:

$$\Phi = \frac{1}{n}\sum_{i=1}^{n}\phi_i \tag{11}$$

Given the value of α, it is called residual qualification model when $\Phi < \alpha$. The value of α can be 0.01, 0.05, or 0.10, and the corresponding model is perfect, qualified, and barely qualified.

As shown in Eq. (12), Gray correlation degree measures the correlating coefficient between the original sequence and the reference sequence:

$$\varepsilon_i(k) = \frac{\min_i \min_k |x(k) - x_i(k)| + \rho \max_i \max_k |x(k) - x_i(k)|}{|x(k) - x_i(k)| + \rho \max_i \max_k |x(k) - x_i(k)|} \tag{12}$$

i denotes the ith group of fitting data, and k denotes the kth one in a certain group. ρ denotes the distinguish coefficient varying from 0 to 1, which is always set as 0.5. However, the correlation coefficient varies with moments, which results in disperse information. Combining the correlation coefficient in different moments together, we can obtain the correlation degree between the original curve and the fitting curve:

$$r_i = \frac{1}{n} \sum_{k=1}^{n} \varepsilon_i(k) \tag{13}$$

2.2 Markov model

Markov chain is proposed by Andrey Markov (1856–1922), and it is a discrete time stochastic process with Markov property in mathematics. Given the current knowledge and information, historical information has no impact on the future. To improve prediction accuracy, Markov model is used to handle the data obtained by GM(1,1). It is critical to divide state and build transition matrix.

2.2.1 Dividing states

To divide states, four rules are suggested to follow. Firstly, the partition state must have at least one true value in each state. Secondly, elements in a one-step transition matrix cannot be the same. Thirdly, the actual values must fall into one state. Finally, the state must pass Markov test. The numbers vary according to the original data. In this chapter, the overall level of FDI in China is on the rise while fluctuating in detail. Therefore, the level of FDI is a non-stable stochastic process. Taking the curve of $\hat{Y}(k) = \hat{x}^{(0)}(k+1)$ as reference, the sequence can be divided into n states. The intervals can be denoted as $Q_i = [Q_{1i}, Q_{2i}]$ and $i = 1, 2, \ldots, n$, in which $Q_{1i} = \hat{Y}(k) + E_{1i}$ and $Q_{2i} = \hat{Y}(k) + E_{2i}$.

2.2.2 Transition matrix

Assuming that there are n states denoting as $E_1, E_2, \ldots, E_n$, the transition probability amounts to frequency approximately in general, namely, $P_{ij} = \frac{M_{ij}^{(l)}}{M_i}$. Then, we can get the lth step transition matrix $P(l) = \left(P_{ij}^{(l)}\right)_{n \times n}$. $M_{ij}^{(l)}$ is the data of raw series transferring l step from the state Q_i to the state Q_j.

2.2.3 The forecasting value

The eventual forecast is in the center of the Gray zone, which is denoted as $Y'(k) = \frac{1}{2}(Q_{1i} + Q_{2i}) = \hat{Y}(k) = \frac{1}{2}(E_{1i} + E_{2i})$. Eventually, the forecasting sequence is obtained as $Y'(k) = \{Y'(1), Y'(2), \ldots, Y'(m)\}$.

3. Time series model (TSM)

Burg suggests that recursive algorithm estimated by the AR(P) model is the most practical one [29], while Hannan proposes time series with multidimensional linear stationary RMA(p, q). The times series model mainly includes the autoregressive model and the moving average model [30–32], and generally the modeling steps are as follows.

3.1 Preliminary analysis of data and modeling identification

Time series prediction is a statistical method processing dynamic data, which is a random sequence arranged in chronological order or a set of ordered random variables defined in probabilistic space $\{X_t, t = 1, 2, \ldots, n\}$, in which the parameter t represents time. In the TSM, if the samples' autocorrelation function $\{\hat{\rho}_k\}$ decreases to zero based on the negative exponential function, then it can be preliminarily judged that this sequence is a stationary autoregressive moving average model (ARMA). If the absolute value of the sample autocorrelation function in the q-step delay $\hat{\rho}_k(k \leq q)$ is greater than twice of the standard deviation and the value of $\hat{\rho}_k(k > q)$ is less than twice of the standard deviation, then the sequence is q-step moving average model (MA(q)). In a similar vein, we can judge p-step autoregressive moving average model (AR(p)) according to the truncation situation of partial autocorrelation function $\{\hat{\varphi}_{kk}\}$.

3.2 Parameter estimation

In order to fit the TSM, we need to estimate the autoregressive coefficient φ_i, the moving average coefficient θ_i, the mean μ, and the variance σ_ε^2 of the white noise sequence in the ARMA model.

3.3 Diagnostic test

The purpose of diagnostic test is to check and test the rationality of the model, including residual test, autocorrelation function of residual error and partial autocorrelation function test, and the significance test of parameters in the model.

3.4 Optimal model selection

Model recognition is only a preliminary selection of TSM. Considering the actual observed errors and statistical errors, several models are taken as candidate models. And the most common methods of selecting optimal models include F-test method, criterion function method (AIC criterion, BIC criterion, SBC criterion).

4. Comparison of GMM and TSM

4.1 GMM predicting FDI of China

Take the FDI value of China over the period from 1990 to 2016 as the original data (unit, \$100 million; data source, Ministry of Commerce of the PRC):

$$X^{(0)} = \{34.87, 43.66, 110.08, \ldots, 1260\}$$

Based on Eq. (5) and using the software MATLAB, the least squares estimation (LSE) of FDI is as follows:

$$\Phi = \begin{pmatrix} \hat{a} \\ \hat{b} \end{pmatrix} = \begin{pmatrix} -0.0697 \\ 243.795 \end{pmatrix}$$

Based on Eq. (7), time-response function can be written as $\hat{x}(k+1) = 3530.59e^{0.0697k} - 3495.72$. Residual values can be obtained according to relative error based on the prediction value of GM(1,1) model. To improve the predicting accuracy, the relative error can be divided into five states (E1, E2, E3, E4, E5) between 1990 and 2010. The relative error status can be seen in **Tables 2**.

According to the original FDI value over a period from 1990 to 2010 and the relative error of prediction value in GM(1,1), the transition matrixes of different steps $P_1^{(i)}(i = 1, 2, 3, 4, 5)$ are shown as follows:

$$P_1^{(1)} = \begin{pmatrix} \frac{3}{5} & \frac{1}{5} & 0 & \frac{1}{5} & 0 \\ 1 & 0 & 0 & 0 & 0 \\ \frac{1}{4} & \frac{1}{4} & 0 & \frac{1}{2} & 0 \\ 0 & 0 & \frac{1}{2} & \frac{1}{2} & 0 \\ 0 & 0 & \frac{1}{2} & 0 & \frac{1}{2} \end{pmatrix}, P_1^{(2)} = \begin{pmatrix} \frac{3}{5} & 0 & 0 & \frac{2}{5} & 0 \\ 1 & 0 & 0 & 0 & 0 \\ \frac{1}{4} & \frac{1}{4} & \frac{1}{4} & \frac{1}{4} & 0 \\ \frac{1}{6} & 0 & \frac{1}{3} & \frac{1}{2} & 0 \\ 0 & \frac{1}{2} & \frac{1}{2} & 0 & 0 \end{pmatrix}, P_1^{(3)} = \begin{pmatrix} \frac{1}{4} & 0 & 0 & \frac{3}{4} & 0 \\ 1 & 0 & 0 & 0 & 0 \\ \frac{1}{2} & 0 & \frac{1}{4} & \frac{1}{4} & 0 \\ \frac{1}{6} & \frac{1}{6} & \frac{1}{3} & \frac{1}{3} & 0 \\ \frac{1}{2} & \frac{1}{2} & 0 & 0 & 0 \end{pmatrix},$$

$$P_1^{(4)} = \begin{pmatrix} 0 & 0 & \frac{1}{4} & \frac{3}{4} & 0 \\ 1 & 0 & 0 & 0 & 0 \\ \frac{2}{3} & 0 & 0 & \frac{1}{3} & 0 \\ \frac{1}{6} & \frac{1}{6} & \frac{1}{3} & \frac{1}{3} & 0 \\ 1 & 0 & 0 & 0 & 0 \end{pmatrix}, P_1^{(5)} = \begin{pmatrix} 0 & 0 & \frac{1}{4} & \frac{3}{4} & 0 \\ 0 & 0 & 0 & 1 & 0 \\ \frac{1}{3} & \frac{1}{3} & \frac{1}{3} & 0 & 0 \\ \frac{2}{5} & 0 & \frac{1}{5} & \frac{2}{5} & 0 \\ 1 & 0 & 0 & 0 & 0 \end{pmatrix}$$

Residual State	E1	E2	E3	E4	E5
Meaning	Extremely underestimated	Underestimated	Reasonable	Overestimated	Extremely overestimated
Range	[−0.17, −0.10]	[−0.10, 0.02]	[0.02, 0.07]	[0.07, 0.12]	[0.12, 0.83]

Table 1.
Relative error status of FDI level in China.

Year	Original	Relative error of GM	State	Year	Original	Relative error of GM	State
1990	34.87	0	E3	2001	468.78	0.0848	E4
1991	43.66	0.8288	E5	2002	527.43	0.0397	E3
1992	110.08	0.5974	E5	2003	535.05	0.0914	E4
1993	275.15	0.0615	E3	2004	606.30	0.0398	E3
1994	337.67	−0.0741	E2	2005	603.25	0.109	E4
1995	375.21	−0.1131	E1	2006	630.21	0.1318	E4
1996	417.26	−0.1545	E1	2007	747.68	0.0394	E3
1997	452.57	−0.1679	E1	2008	923.95	−0.1071	E1
1998	454.63	−0.0941	E1	2009	900.33	−0.0061	E2
1999	403.19	0.095	E4	2010	1057.35	−0.102	E1
2000	407.15	0.1477	E4	—	—	—	—

Data source: China Statistical Yearbook over the period from 2000 to 2006, Ministry of Commerce of the PRC.

Table 2.
Comparison of GM(1,1) prediction value and original value of FDI of China.

Based on the transition matrix, we can obtain the error state over a period from 2011 to 2016 (see **Table 3**). Taking the middle value of the error state to modify the prediction value of GM(1,1) model, then the modified value can be seen in **Table 3**. And $x^{(0)}(k)$, $\hat{x}^{(0)}(k)$, and $\phi(i)$ represent the original value, predicting value and relative error of GM(1,1). $\hat{x'}^{(0)}(k)$ and $\phi'(i)$ represent the modified value and relative error of GMM.

In the light of Eqs. (9)–(11), the relative residual error of GM(1,1) and GMM is $0.0584 < \alpha = 0.1$ and $0.0458 < \alpha = 0.05$, respectively. Therefore, the GM(1,1) model is barely qualified, and the modified GMM model is qualified. Gray correlation degrees of the two models are 67 and 79.9%, respectively. In summary, the

Year	$x^{(0)}(k)$	$\hat{x}^{(0)}(k)$	$\phi(i)$	$\hat{x'}^{(0)}(k)$	$\phi'(i)$
1990	34.87	0.0349	0	33.4752	−0.0471
1991	43.66	0.2550	0.8288	127.5082	0.6739
1992	110.08	0.2734	0.5974	136.7182	0.2332
1993	275.15	0.2932	0.0615	281.4594	0.0173
1994	337.67	0.3144	−0.0741	326.9385	−0.0328
1995	375.21	0.3371	−0.1131	379.20437	0.0193
1996	417.26	0.3614	−0.1545	406.5945	−0.0172
1997	452.57	0.3875	−0.1679	435.9630	−0.0289
1998	454.63	0.4155	−0.0941	467.4528	0.0360
1999	403.19	0.4455	0.0950	392.0632	0.0000
2000	407.15	0.4777	0.1477	420.3821	0.0582
2001	468.78	0.5122	0.0848	450.7465	−0.0113
2002	527.43	0.5492	0.0397	527.2409	−0.0056
2003	535.05	0.5889	0.0914	518.2134	−0.0040
2004	606.30	0.6314	0.0398	606.1574	−0.0055
2005	603.25	0.6770	0.1090	595.7787	0.0154
2006	630.21	0.7259	0.1318	638.8121	0.0407
2007	747.68	0.7784	0.0394	747.2223	−0.0059
2008	923.95	0.8346	−0.1071	938.8998	0.0246
2009	900.33	0.8949	−0.0061	930.6540	0.0326
2010	1057.35	0.9595	−0.1020	1079.4327	0.0291
2011	1160.11	1.0288	−0.1276	1157.4006	0.0065
2012	1117.20	1.1031	−0.0128	1241.0002	0.1077
2013	1175.90	1.1828	0.0058	1330.6383	0.1241
2014	1195.60	1.2682	0.0573	1116.0363	−0.0417
2015	1262.70	1.3598	0.0714	1196.648	−0.0260
2016	1260.00	1.4580	0.1358	1283.0826	0.0451

Annotations: The unit of $x^{(0)}(k)$ and $\hat{x'}^{(0)}(k)$ is 1 billion dollars. The unit of $\hat{x}^{(0)}(k)$ is 10^3 billion dollars. Data source: China Statistical Yearbook.

Table 3.
Residual checklist of Markov model and GM(1,1).

prediction accuracy of GMM has been improved, and its fitting effect exceeds the model of GM(1,1).

4.2 TSM predicting FDI of China

Now we will build a TSM based on the FDI value of China over the period from 1990 to 2016, obtain the predicting data, compare the difference between the predicted data and the original date, and evaluate the accuracy of this model.

Figure 1 shows the changing tendency of FDI in China over the period between 1990 and 2016. The raw data series show the seasonal change and overall growth, but the data series are not stable. Through the seasonal difference method to process the data, the seasonal difference order of three was selected. After the differential processing, the data sequence has been stabilized, eliminating the growing trend (**Figure 2**).

We determine the order of TSM based on sample autocorrelation function and partial autocorrelation function. After the one-step delay, the sample autocorrelation function falls to a standard error of twice times and has the property of truncation. After the two-step delay, the sample partial autocorrelation function falls to a standard error of twice times and has the property of truncation.

In the light of the calculation of SAS software, now we compare the model of ARMA(2,1), AR(2), and MA(1) (see **Tables 4** and **5**).

Comparing the AIC and SBC values for ARMA(2,1), AR(2), and MA(1) models (see **Table 4**), we find the model MA(1) to be the most inferior. Considering the AIC and SBC criterion values of ARMA(2,1) and AR(2) and the significance of parameters, it is found that fitting effect of the AR(2) model is the best.

As shown in **Table 5**, the P-value (Pr > ChiSq) for self-correlation test of the residual sequence with the 6-step delay, the 12-step delay and the 18-step delay are greater than that of the significant level $\alpha = 0.1$. Therefore, we cannot reject the hypothesis that residuals are non-autocorrelated. That is to say, the residual is regarded as a white noise sequence. This illustrates that the AR(2) model has extracted sufficient information from the raw series and it is a rational model:

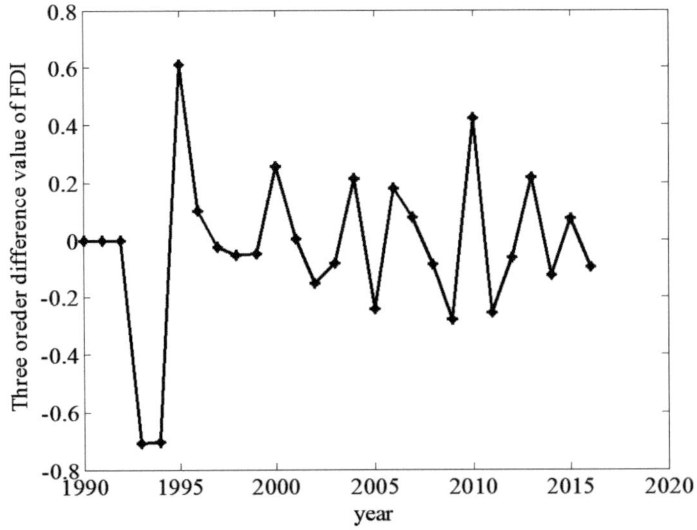

Figure 2.
The curve about time after differential.

Model	Parameter	Estimate	P-value	AIC	SBC
AR(2)	MU	2.0711	<0.0001***	1.0603	4.5944
	AR1,1	1.5357	<0.0001***		
	AR1,2	−0.5392	0.0102**		
MA(1)	MU	0.4676	0.0038***	28.7297	31.08558
	MA1,1	−0.7099	0.0001***		
ARMA(2,1)	MU	1.9380	<0.0001***	0.7062	5.4184
	MA1,1	−0.4940	0.1192		
	AR1,1	1.2352	0.0015***		
	AR1,2	−0.2358	0.5028		

*Annotations: ***, **, and * indicate a significant level of 0.01, 0.05, and 0.1, respectively.*

Table 4.
Prediction results of TSM.

To lag	6	12	18
Chi-square	3.91	5.44	8.02
Pr > ChiSq	0.4187	0.8599	0.9481

Table 5.
Self-correlation test of AR(2) model.

$$(1 - 1.53571B + 0.53921B^2)(1 - B^3)X_t = \varepsilon_t$$

where $X = num - 2.0711$, $t = year$, and num represents the FDI value of the corresponding year.

4.3 Comparison of prediction results of two models

4.3.1 Accuracy assessment

Regarding how to select the appropriate accuracy evaluation criteria, Yokuma and Armstrong [33] have done a survey of expert opinions. They think that accuracy, clear physical meaning, and being easy to implement can be the critical evaluation criteria [33]. Accordingly, three criteria are used to evaluate the accuracy of the prediction model.

Criterion	Mean squared error	Mean absolute error	Mean absolute percentage error
Index	$MSE = \frac{1}{n}\sum_{i=1}^{n}(x_i - \hat{x}_i)^2$	$MAE = \frac{1}{n}\sum_{i=1}^{n}\lvert x_i - \hat{x}_i\rvert$	$MAPE = \frac{1}{n}\sum_{i=1}^{n}\left\lvert\frac{x_i - \hat{x}_i}{x_i}\right\rvert$
GM(1,1)	7.3991e+03	66.0812	0.3117
GMM	2.4731e+03	29.5558	0.1181
Time series	1.6644e+04	85.6101	0.1448

Annotations: $\hat{x}_i$ is the predicting value, x_i is the original value, and n is the predicting number.

Table 6.
Three criteria to evaluate the accuracy of models.

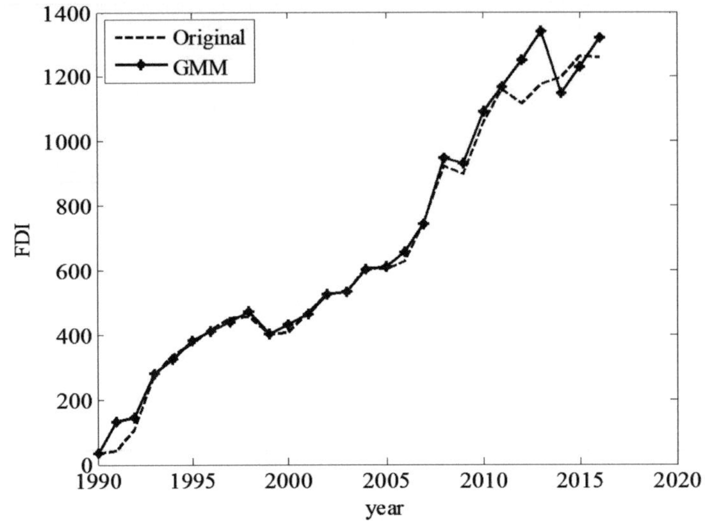

Figure 3.
Original value and that in GMM.

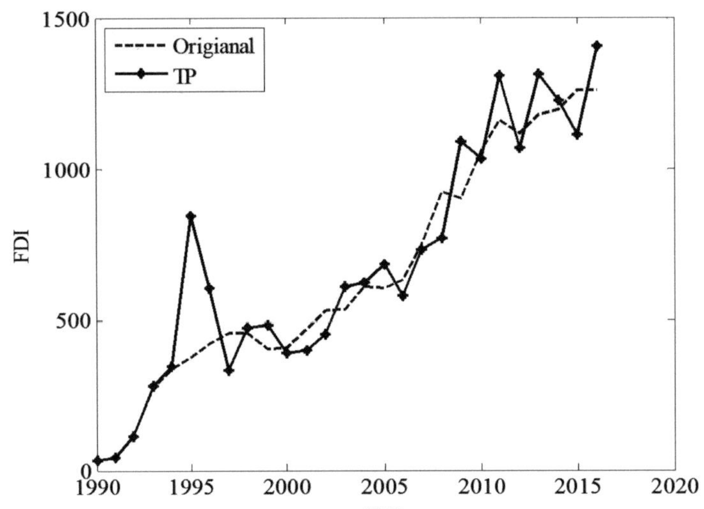

Figure 4.
Original value and that in TSM.

4.3.2 Comparing predicted values with actual values

As shown in **Table 6**, the prediction accuracy of GMM has been improved manifestly compared with that in GM(1,1) model. Therefore, the forecasting value in GMM is closer to the actual level of China's FDI. Then, from **Figures 3** and **4**, we can clearly see that GMM model has a better fitting effect than that in TSM.

5. Empirical analysis of FDI in Chongqing and Beijing

From discussions above, it is found that GMM has higher prediction accuracy and better fitting effects than those of TSM of Chinese FDI level. To further verify the credibility of this result, we construct GMM and TSM based on the FDI level of

Beijing (1990–2016) and Chongqing (1990–2015). The divided states involved in the GMM are shown in **Table 7**, and the transition matrixes of GMM associated with Beijing and Chongqing are denoted as P_2 and P_3. For simplification, we only list the form of transition matrix P_2. The comparison of GM(1,1) and GMM can be seen in **Tables 8** and **9**. The average relative errors of GM(1,1) and GMM of Beijing (Chongqing) are 0.0312 (0.5285) and −0.0029 (−0.1051), respectively. The Gray

Area	Error state	E1	E2	E3	E4	E5
Beijing	Range	[−0.47, −0.2]	[−0.2, −0.1]	[−0.1, 0.1]	[0.1, 0.28]	[0.28, 0.65]
Chongqing	Range	[0, 0.28]	[0.28, 0.55]	[0.55, 0.75]	[0.75, 0.81]	[0.81,0.97]

Table 7.
Residual states of FDI in Beijing and Chongqing.

Year	Original value	GM(1,1)	GMM	TSM	To state	GR	MR
1990	27,696	0.0277	0.0277	0.0277	E3	0	0
1991	24,482	0.0693	0.0371	0.0245	E5	0.6466	0.3395
1992	34,984	0.0780	0.0417	0.0364	E5	0.5514	0.1614
1993	66,693	0.0878	0.0711	0.0311	E4	0.2401	0.0618
1994	144,460	0.0988	0.1121	0.0863	E1	−0.4625	−0.2885
1995	140,277	0.1112	0.1262	0.1342	E1	−0.2617	−0.1117
1996	155,290	0.1251	0.1420	0.1969	E1	−0.2410	−0.0934
1997	159,286	0.1408	0.1620	0.1514	E2	−0.1310	0.0165
1998	216,800	0.1585	0.1799	0.2127	E1	−0.3677	−0.2050
1999	197,525	0.1784	0.2052	0.2126	E2	−0.1071	0.0373
2000	168,368	0.2008	0.1626	0.2680	E4	0.1615	−0.0352
2001	176,818	0.2260	0.1831	0.1767	E4	0.2176	0.0341
2002	172,464	0.2544	0.1361	0.2213	E5	0.3220	−0.2673
2003	219,126	0.2863	0.2319	0.1890	E4	0.2346	0.0551
2004	255,974	0.3222	0.2610	0.2560	E4	0.2056	0.0193
2005	352,834	0.3627	0.3627	0.2878	E3	0.0271	0.0271
2006	455,191	0.4082	0.4694	0.3979	E2	−0.1151	0.0303
2007	506,572	0.4594	0.5283	0.5179	E2	−0.1026	0.0412
2008	608,172	0.5171	0.5947	0.5870	E2	−0.1761	−0.0227
2009	612,094	0.5820	0.5820	0.6851	E3	−0.0517	−0.0517
2010	636,358	0.6550	0.6550	0.7277	E3	0.0285	0.0285
2011	705,447	0.7373	0.8368	0.7195	E5	0.0431	0.1570
2012	804,160	0.8298	0.6721	0.8222	E4	0.0309	−0.1964
2013	852,418	0.9339	0.7565	0.9097	E4	0.0873	−0.1268
2014	904,085	1.0512	1.0512	1.0009	E3	0.1399	0.1399
2015	1,299,635	1.1831	1.3428	1.0292	E1	−0.0985	0.0322
2016	1,302,858	1.3316	1.5114	1.4400	E1	0.0216	0.1380

*Annotations: In **Table 8**, the unit of original value is 1 million dollars. GM, GMM, and TSM represent the predicted value of Gray model, Gray-Markov model, and time series model, and their unit is 10^6 million dollars. GR and MR represent the residuals of Gray model and Gray-Markov model. Data source: Beijing Statistical Yearbook (1990–2017), Beijing Municipal Bureau of Statistics.*

Table 8.
Comparison of predicted errors of GMM and GM(1,1) of Beijing FDI level.

relational degrees of GM(1,1) and GMM of Beijing (Chongqing) are 64.62% (75.26%) and 79.39% (86.82%), respectively. Therefore, the errors of GM(1,1) and GMM are barely qualified or qualified, and hence GMM is superior to GM(1,1):

$$P_1^{(1)} = \begin{pmatrix} \frac{2}{4} & \frac{4}{4} & 0 & 0 & 0 \\ \frac{1}{5} & \frac{2}{5} & \frac{1}{5} & \frac{1}{5} & 0 \\ 0 & \frac{1}{3} & \frac{1}{3} & 0 & \frac{1}{3} \\ \frac{1}{5} & 0 & \frac{1}{5} & \frac{2}{5} & \frac{1}{5} \\ 0 & 0 & 0 & \frac{2}{3} & \frac{1}{3} \end{pmatrix}, P_1^{(2)} = \begin{pmatrix} \frac{2}{4} & \frac{1}{4} & 0 & \frac{1}{4} & 0 \\ 0 & \frac{2}{5} & \frac{2}{5} & \frac{1}{5} & 0 \\ 0 & \frac{1}{2} & 0 & 0 & \frac{1}{2} \\ \frac{1}{5} & \frac{1}{5} & \frac{1}{5} & \frac{1}{5} & \frac{1}{5} \\ \frac{1}{3} & 0 & 0 & \frac{2}{3} & 0 \end{pmatrix}, P_1^{(3)} = \begin{pmatrix} \frac{1}{4} & \frac{2}{4} & 0 & \frac{1}{4} & 0 \\ 0 & 0 & \frac{2}{4} & \frac{1}{4} & \frac{1}{4} \\ 0 & \frac{1}{2} & 0 & \frac{1}{2} & 0 \\ \frac{1}{5} & \frac{2}{5} & 0 & \frac{2}{5} & 0 \\ \frac{2}{3} & 0 & \frac{1}{3} & 0 & 0 \end{pmatrix},$$

$$P_1^{(4)} = \begin{pmatrix} \frac{1}{4} & \frac{1}{4} & \frac{1}{4} & 0 & \frac{1}{4} \\ 0 & 0 & \frac{1}{3} & \frac{2}{3} & 0 \\ \frac{1}{2} & 0 & \frac{1}{2} & 0 & 0 \\ 0 & \frac{3}{5} & \frac{1}{5} & \frac{1}{5} & 0 \\ \frac{2}{3} & \frac{1}{3} & 0 & 0 & 0 \end{pmatrix}, P_1^{(5)} = \begin{pmatrix} 0 & \frac{1}{4} & 0 & \frac{3}{4} & 0 \\ 0 & 0 & 0 & \frac{1}{2} & \frac{1}{2} \\ \frac{1}{2} & 0 & \frac{1}{2} & 0 & 0 \\ \frac{1}{5} & \frac{2}{5} & \frac{2}{5} & 0 & 0 \\ \frac{1}{3} & \frac{2}{3} & 0 & 0 & 0 \end{pmatrix}$$

Year	Original value	GM(1,1)	GMM	TSM	To state	GR	MR
1990	332	0.0033	0.0029	0.0003	E1	0	−0.1628
1991	977	0.3159	0.0347	0.0010	E5	0.9691	0.7188
1992	10,247	0.3647	0.1276	0.0029	E3	0.7190	0.1972
1993	25,915	0.4210	0.2463	0.0846	E2	0.3844	−0.0523
1994	44,953	0.4860	0.4180	0.0686	E1	0.0751	−0.0755
1995	37,926	0.5611	0.3282	0.0842	E2	0.3241	−0.1554
1996	21,878	0.6478	0.2267	0.0406	E3	0.6623	0.0350
1997	38,466	0.7478	0.4375	0.0165	E2	0.4856	0.1207
1998	43,107	0.8633	0.5050	0.0755	E2	0.5007	0.1465
1999	23,893	0.9967	0.2193	0.0563	E4	0.7603	−0.0897
2000	24,436	1.1506	0.2531	0.0181	E4	0.7876	0.0347
2001	25,649	1.3284	0.2922	0.0296	E4	0.8069	0.1223
2002	28,089	1.5335	0.1687	0.0329	E5	0.8168	−0.6651
2003	31,112	1.7704	0.1947	0.0360	E5	0.8243	−0.5976
2004	40,508	2.0439	0.4497	0.0417	E4	0.8018	0.0991
2005	51,575	2.3596	0.5191	0.0599	E4	0.7814	0.0065
2006	69,595	2.7241	0.9534	0.0776	E3	0.7445	0.2700
2007	108,534	3.1448	1.1007	0.1059	E3	0.6549	0.0139
2008	272,913	3.6306	3.1223	0.1930	E1	0.2483	0.1259

Year	Original value	GM(1,1)	GMM	TSM	To state	GR	MR
2009	401,643	4.1914	3.6046	0.6941	E1	0.0417	−0.1143
2010	304,264	4.8388	2.8307	0.6809	E2	0.3712	−0.0749
2011	582,575	5.5862	3.2679	0.2878	E2	−0.0429	−0.7827
2012	352,418	6.4490	3.7727	1.2269	E2	0.4535	0.0659
2013	414,353	7.4452	4.3554	0.2769	E2	0.4435	0.0487
2014	423,348	8.5952	1.8909	0.5837	E4	0.5075	−1.2388
2015	377,183	9.9228	2.1830	0.5124	E4	0.6199	−0.7278
2016	332	0.0033	0.0029	0.0003	E1	0	−0.1628

*Annotations: In **Table 9**, the unit of original value is 1 million dollars. GM, GMM, and TSM represent the predicted value of Gray model, Gray-Markov model, and time series model, and their unit is 10^5 million dollars. GR and MR represent the residuals of Gray Model and Gray-Markov model. Data source: Chongqing Statistical Yearbook (1990–2016), Beijing Municipal Bureau of Statistics.*

Table 9.
Comparison of predicted errors of GMM and GM(1,1) of Chongqing FDI level.

Similar to Section 4.2, TSM of Beijing FDI can be modeled as MA (1):

$$(1 + 0.82673B)(1 - B^2)X_t = \varepsilon_t$$

where $X = num - 0.27264$ and $t = year$.
TSM of Chongqing FDI can be modeled as ARMA(1,2,1):

$$(1 - 0.82442B)(1 + B)(1 - B^2)X_t = \varepsilon_t$$

where $X = num - 2.178452$ and $t = year$.
Figure 5 (**Figure 6**) shows the difference between the original value and the predicting value in Gray-Markov model (time series model) of foreign direct investment in Beijing. It is apparent that the fitting effect of GMM is better than

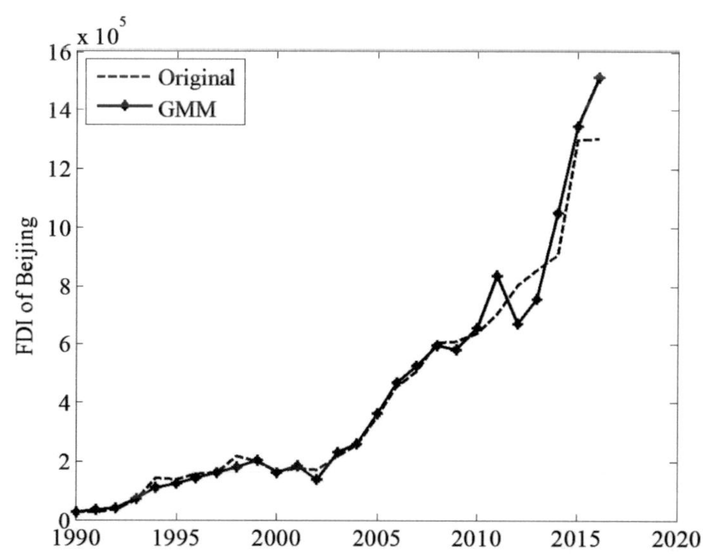

Figure 5.
Original value and that in GMM of BJ. Annotations: BJ denotes the city of Beijing.

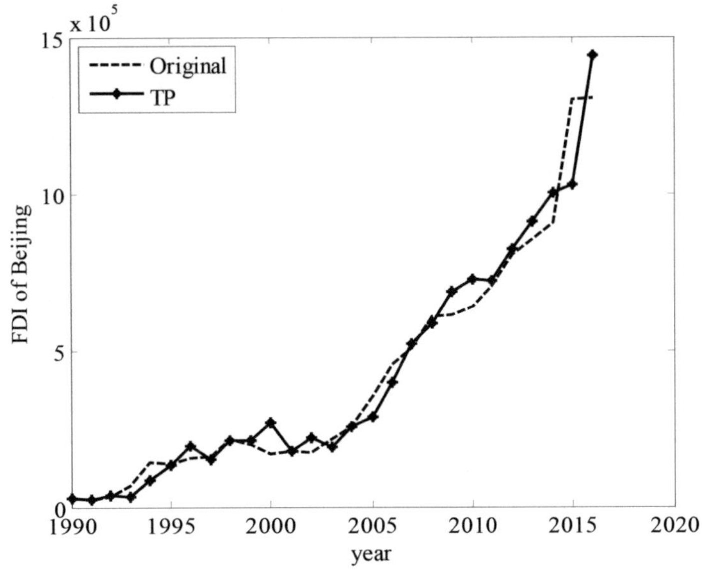

Figure 6.
Original value and that in TSM of BJ. Annotations: BJ denotes the city of Beijing.

that of TSM. The similar conclusion can be drawn from **Figures 7** and **8**. **Tables 9** and **10** show the predicting effect of GMM is better than that of TSM from the point of predicting errors and accuracy. There is no doubt that it is a good thing to predict accurately the foreign direct investment of the forthcoming 5 or 10 years for the domain specialists. Because if the predicting results is lower or higher than they expected, they could pay attention to seeking the critical factors and policy which have impacts on FDI and adjust them in advance.

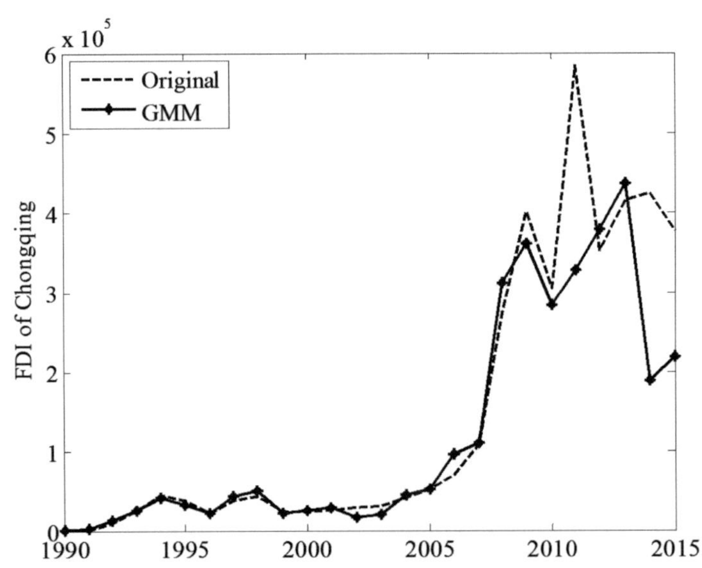

Figure 7.
Original value and that in GMM of CQ. Annotations: CQ denotes the city of Chongqing.

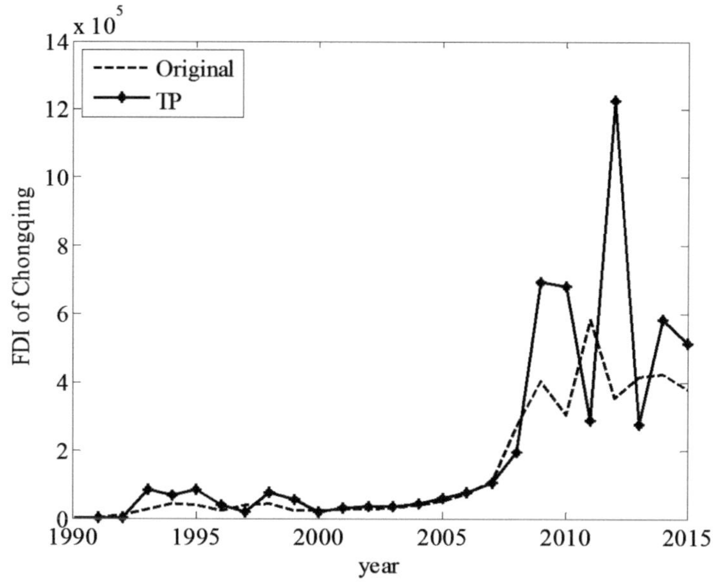

Figure 8.
Original value and that in TSM of CQ. Annotations: CQ denotes the city of Chongqing.

Area	Index	$MSE = \frac{1}{n}\sum_{i=1}^{n}(x_i - \hat{x}_i)^2$	$MAE = \frac{1}{n}\sum_{i=1}^{n}\lvert x_i - \hat{x}_i\rvert$	$MAPE = \frac{1}{n}\sum_{i=1}^{n}\left\lvert\frac{x_i - \hat{x}_i}{x_i}\right\rvert$
Beijing	GM(1,1)	3.3244e+09	4.7473e+04	0.2587
	GMM	4.3599e+09	3.9369e+04	0.1032
	TSM	5.4378e+09	4.6731e+04	0.1335
Chongqing	GM(1,1)	3.9173e+10	1.3478e+05	2.9584
	GMM	5.8230e+09	3.4080e+04	0.2663
	TSM	4.4524e+10	1.0126e+05	0.6018

Annotations: MSE, MAE, and MAPE denote mean squared error, mean absolute error, and mean absolute percentage error.

Table 10.
Comparison of predicting accuracy of two models.

6. Conclusions and future work

Our contributions are threefold. Firstly, comparing the predicting results of the Gray-Markov model and the time series model and the original value, respectively, we can find that the fitting effect of the former (GMM) is better than the latter (TSM) and so does its scientific and practical importance. Secondly, the predicting results of GMM show that the level of foreign investment in China has been increasing by years. Thirdly, in order to further enhance Chinese international status and attract more foreign investment, the government should play a role at a macro level to reduce excessive market administrative intervention, establish a service-oriented government, and reduce the relevant approval procedures for international investment.

In the future work, the Gray-Markov model and time series model can be combined with other predicting model (e.g., support vector machine and dynamic Bayesian) to improve the accuracy. Also these models have the potential to be applied in the other areas such as finance (e.g., stocks, funds, and security), risk (e.g., financial risk and operational risk), and business (e.g., consumer price index and incomes).

Acknowledgements

This chapter is financially supported by the National Natural Science Foundation of China under grant nos. 71771080, 71172194, 71521061, 71790593, 71642006, 71473155, 71390335, and 71571065.

Author details

Yanyan Zheng[1*], Tong Shu[2], Shou Chen[2] and Kin Keung Lai[3,4]

1 School of Management, Xi'an Polytechnic University, Xi'an, China

2 Business School, Hunan University, Changsha, China

3 International Business School, Shaanxi Normal University, Xi'an, China

4 Department of Management Sciences, City University of Hong Kong, Kowloon, Hong Kong, China

*Address all correspondence to: yanzheng@whu.edu.cn; shutong@hnu.edu.cn

References

[1] Choe JI. Do foreign direct investment and gross domestic investment promote economic growth. Review of Development Economics. 2003;**7**(1): 44-57

[2] Shu T, Liu CX, Chen S, Wang SY, Li JQ. The influence and demonstration of FDI on China's economic growth in supply chain system. Systems Engineering - Theory & Practice. 2014; **34**(2):282-290

[3] Wei KL. Foreign direct investment and economic growth in China. Beijing: Economic Science Press; 2013

[4] Pan FH, Wang JC. From "passive embedding" to supply Chain Park investment: A new model of FDI. China Soft Science. 2010;**25**(3):95-102

[5] Dees S. Foreign direct investment in China: Determinants and effects. Economics of Planning. 1998;**31**(2): 175-194

[6] Nourzad F. Openness and Efficiency of FDI: A panel stochastic production frontier study. International Advances in Economic Research. 2008;**14**(1):25-35

[7] Mah JS. Foreign direct investment inflows and economic growth of China. Economic Research Journal. 2002;**48**(3): 69-75

[8] Pan WR, Song Y. Empirical analysis for the impact of RMB real effective exchange rate on foreign direct investment in China. Journal of Chemical and Pharmaceutical Research. 2014;**6**(5):1830-1836

[9] De Mello LR Jr. Foreign direct in developing countries and growth: A selective survey. The Journal of Development Studies. 1997;**34**(1):1-34

[10] Dreher A, Mikosch H, Voigt S. Membership has its privileges-The effect of membership in international organizations on FDI. World Development. 2015;**66**:346-358

[11] Badr OM, Ayed TL. The mediator role of FDI in North Africa: Case of Egypt. 2015;**3**(1):1-7

[12] Kathuria V, Ray P, Bhangaonkar R. FDI (foreign direct investment) in wind energy sector in India: Testing the effectiveness of state policies using panel data. Energy. 2015;**80**:190-202

[13] Lin HL, Hsiao YC, Lin ES. The choice between standard and non-standard FDI production strategies for Taiwanese multinationals. Research Policy. 2015;**44**(1):283-293

[14] Brülhart M, Schmidheiny K. Estimate the rivalness of state-level inward FDI. Journal of Regional Science. 2015;**55**(1):139-148

[15] Wang YB, Papageorgiou M. Real-time freeway traffic state estimation based on extended Kalman filter: A general approach. Transportation Research Part B. 2005;**39**(2):141-167

[16] Bhattacharya PS, Thomakos DD. Forecasting industry-level CPI and PPI inflation: Does exchange rate pass-through matter? International Journal of Forecasting. 2008;**24**(1):134-150

[17] Li GD, Yamaguchi D, Nagai M. Application GM(1,1)-Markov chain combined model to China automobile industry. International Journal of Industrial and Systems Engineering. 2007;**2**(3):327-347

[18] Wang CB, Li LH. The forecast of gold price based on the GM(1,1) and Markov chain. Gray Systems and Intelligent Services. 2007;**18**(1):739-743

[19] Lin CT, Lee IF. Artificial intelligence diagnosis algorithm for expanding a

precision expert forecasting system. Expert Systems with Application. 2009; **36**(4):8385-8390

[20] Aries MB, Veitch JA, Newsham GR. Windows, view, and office characteristics predict physical and psychological discomfort. Journal of Environmental Psychology. 2010;**30**(4): 533-541

[21] Li GD, Masuda S, Yamaguchi D, Nagai M, Wang CH. An improved grey dynamic GM(2,1) model. International Journal of Computer Mathematics. 2010;**87**(7):1617-1629

[22] Li GD, Masuda S, Nagai M. The prediction model for electrical power system using an improved hybrid optimization model. Electrical Power and Energy Systems. 2013;**44**(1): 981-987

[23] Lin LC, Wu SY. Analyzing Taiwan IC assembly industry by Gray-Markov fore casting model. Mathematical Problems in Engineering. 2013;**70**(2): 717-718

[24] Chen YY, Liu CX. An improved GM (1,1)-Markov model in supply chain disruption and its application in demand prediction. Information Technology Journal. 2014;**13**(13):2204-2210

[25] Dong S, Chi K, Zhang QY, Zhang XD. The application of a Gray Markov model to forecasting annual maximum water levels at hydrological stations. Journal of Ocean University of China. 2012;**11**(1):13-17

[26] Syamala P, Vishnu B. Gray model for stream flow prediction. Aceh International Journal of Science and Technology. 2012;**1**(1):14-19

[27] Sun W, Wang JM, Chang H. Forecasting carbon dioxide emissions in china using optimization grey model. Journal of Computers. 2013;**8**(1):97

[28] He Y, Bao YD. Gray Markov model and its application. Systems Engineering - Theory & Practice. 1992;**12**(4):59-63

[29] Burg JP. Maximum Entropy Spectral Analysis. In: Proceedings of the 37th Meeting of the Society of Exploration Geophysicists, Oklahoma City, Oklahoma; 1967

[30] Wang ZX, Chan LW. Learning causal relations in multivariate time series data. ACM Transactions on Intelligent Systems and Technology. 2012;**3**(4):76-104

[31] Kadri F, Harrou F, Chaabane S, Tahon C. Time series modelling and forecasting of emergency department overcrowding. Journal of Medical Systems. 2014;**38**(9):1-20

[32] Rivero CR, Pucheta J, Laboret S. Time series forecasting using bayesian method: Application to cumulative rainfall. IEEE Latin America Transactions. 2013;**11**(1):359-364

[33] Yokum JT, Armstrong JS. Beyond accuracy: Comparison of criteria used to select forecasting methods. International Journal of Forecasting. 1995;**11**(4):591-597